W9-BTL-036

Green
Metropolis

ALSO BY DAVID OWEN

Sheetrock & Shellac (2006)

Copies in Seconds (2004)

The First National Bank of Dad (2003)

Hit & Hope (2003)

The Chosen One (2001)

The Making of the Masters (1999)

Around the House
(also published as *Life Under a Leaky Roof*; 1998/2000)

Lure of the Links (coeditor; 1997)

My Usual Game (1995)

The Walls Around Us (1991)

The Man Who Invented Saturday Morning (1988)

None of the Above (1985)

High School (1981)

Riverhead Books   a member of Penguin Group (USA) Inc.   New York   2009

Green
Metropolis

Why Living Smaller, Living Closer, and
Driving Less Are the Keys to Sustainability

DAVID OWEN

RIVERHEAD BOOKS
Published by the Penguin Group
Penguin Group (USA) Inc., 375 Hudson Street, New York, New York 10014, USA •
Penguin Group (Canada), 90 Eglinton Avenue East, Suite 700, Toronto, Ontario M4P 2Y3,
Canada (a division of Pearson Canada Inc.) • Penguin Books Ltd, 80 Strand,
London WC2R 0RL, England • Penguin Ireland, 25 St Stephen's Green, Dublin 2, Ireland
(a division of Penguin Books Ltd) • Penguin Group (Australia), 250 Camberwell Road,
Camberwell, Victoria 3124, Australia • (a division of Pearson Australia Group Pty Ltd) •
Penguin Books India Pvt Ltd, 11 Community Centre, Panchsheel Park, New Delhi–110 017,
India • Penguin Group (NZ), 67 Apollo Drive, Rosedale, North Shore 0632, New Zealand
(a division of Pearson New Zealand Ltd) • Penguin Books (South Africa) (Pty) Ltd,
24 Sturdee Avenue, Rosebank, Johannesburg 2196, South Africa

Penguin Books Ltd, Registered Offices: 80 Strand, London WC2R 0RL, England

Copyright © 2009 by David Owen
All rights reserved. No part of this book may be reproduced, scanned, or distributed in any
printed or electronic form without permission. Please do not participate in or encourage piracy
of copyrighted materials in violation of the author's rights. Purchase only authorized editions.
Published simultaneously in Canada

Parts of several chapters of this book first appeared, in different form, in *The New Yorker.*
Part of chapter 6 first appeared, in different form, in *Golf Digest Index.*

Library of Congress Cataloging-in-Publication Data

Owen, David.
    Green metropolis ; why living smaller, living closer, and driving less are the keys to sustainability
        p.   cm.
    Includes bibliographical references and index.
    ISBN 978-1-59448-882-5
    1. Human ecology—New York (State)—New York.   2. Urban ecology (Sociology)—New York
(State)—New York.   3. Social ecology—New York (State)—New York.   4. Sustainable living—
New York (State)—New York.   5. Sustainable architecture—New York (State)—New York.
6. Green technology—New York (State)—New York.   7. New York (N.Y.)—Environmental
conditions.   I. Title.
    GF504.N7O95   2009   2009017116
    304.2091732—dc22

Printed in the United States of America
10   9   8   7   6   5   4   3   2   1

Book design by Meighan Cavanaugh

While the author has made every effort to provide accurate telephone numbers
and Internet addresses at the time of publication, neither the publisher nor the
author assumes any responsibility for errors, or for changes that occur after publication.
Further, the publisher does not have any control over and does not assume any responsibility
for author or third-party websites or their content.

CONTENTS

Green
Metropolis

One

More Like Manhattan

My wife and I got married right out of college, in 1978. We were young and naïve and unashamedly idealistic, and we decided to make our first home in a utopian environmentalist community in New York state. For seven years we lived quite contentedly in circumstances that would strike most Americans as austere in the extreme: our living space measured just seven hundred square feet, and we didn't have a lawn, a clothes dryer, or a car. We did our grocery shopping on foot, and when we needed to travel longer distances we used public transportation. Because space at home was scarce, we seldom acquired new possessions of significant size. Our electric bill worked out to about a dollar a day.

The utopian community was Manhattan. Most Americans, including most New Yorkers, think of New York City as an ecological nightmare, a wasteland of concrete and garbage and diesel fumes and traffic jams, but in comparison with the rest of America it's a model of environmental responsibility. In fact, by the most significant measures, New York is the greenest community in the United States. The most devastating damage that humans have done to the environment has arisen from the burning of fossil fuels, a category in which New Yorkers are practically prehistoric by comparison with other Americans, including people who live in rural areas or in such putatively eco-friendly cities as Portland, Oregon, and Boulder, Colorado. The average Manhattanite consumes gasoline at a rate that the country as a whole hasn't matched since the mid-1920s, when the most widely owned car in the United States was the Ford Model T.[1] Thanks to New York City, the average resident of New York state uses less gasoline than the average resident of any other state, and uses less than half as much as the average resident of Wyoming. Eighty-two percent of employed Manhattan residents travel to work by public transit, by bicycle, or on foot. That's ten times the rate for Americans in general, and eight times the rate for workers in Los Angeles County.[2] New York City is more populous than all but eleven states; if it were granted statehood, it would rank fifty-first in per-capita energy use, not only because New Yorkers drive less but because city dwellings are smaller than other American dwellings and are less likely to contain a superfluity of large appliances.[3] The average New Yorker (if one takes into consideration all five boroughs of the city) annually generates 7.1 metric tons of greenhouse gases, a lower rate than that of residents of any other American city, and less than 30 percent of the national average, which is 24.5 metric tons[4]; Manhattanites generate even less.

"Anyplace that has such tall buildings and heavy traffic is obviously an environmental disaster—except that it isn't," John Holtzclaw, who recently retired as the chairman of the Sierra Club's transportation committee, told me in 2004. "If New Yorkers lived at the typical American sprawl density of three households per residential acre, they would require many times as much land. They'd be driving cars, and they'd have huge lawns and be using pesticides and fertilizers on them, and then they'd be overwatering their lawns, so that runoff would go into streams." The key to New York's relative environmental benignity is its extreme compactness. Charles Komanoff, a New York City economist, environmental activist, and bicycling enthusiast, told me, "New Yorkers trade the supposed convenience of the automobile for the true convenience of proximity. They are able to live without the ecological disaster of cars—which is caused not just by having to use a car for practically every trip, but also by the distance that you have to traverse. Bicycling, transit, and walking support each other, because they are all made possible by population density." Manhattan's density is approximately 67,000 people per square mile, or more than eight hundred times that of the nation as a whole and roughly thirty times that of Los Angeles. Placing one and a half million people on a twenty-three-square-mile island sharply reduces their opportunities to be wasteful, enables most of them to get by without owning cars, encourages them to keep their families small, and forces the majority to live in some of the most inherently energy-efficient residential structures in the world: apartment buildings. It also frees huge tracts of land for the rest of America to sprawl into.

My wife, whose name is Ann Hodgman, and I had our first child, Laura, in 1984. Ann and I had grown up in suburbs, and we decided that we didn't want to raise Laura in a huge city. A couple of months after she learned to walk, we moved to a small town in the northwest corner of Connecticut, about ninety miles north of midtown Manhattan. Our house was built in the late 1700s. During a rainstorm one night soon after we moved in, I stuck my head into the attic and ran a flashlight over the underside of the roof. The decking boards had been made, two hundred years before, from the broad trunks of old-growth American chestnut trees, a species that was wiped out by an imported blight in the first half of the twentieth century, and some of them were almost as broad as sheets of plywood. The rafters, which were hand-hewn, were joined not by iron nails but by wooden pegs. Carved near the ends of some of the rafters were large Roman numerals, which had been placed there as assembly aids by the anonymous eighteenth-century builder. The house is across a dirt road from a nature preserve and is shaded by tall white-pine trees, and after the storm had ended I could hear a swollen creek rushing past at the bottom of the hill. Deer, wild turkeys, and the occasional black bear feed themselves in our yard, and wildflowers grow everywhere. From the end of our driveway, I can walk several miles through woods to an abandoned nineteenth-century railway tunnel, while crossing only one paved road.

Yet our move was an ecological catastrophe. Our consumption of electricity went from roughly 4,000 kilowatt-hours a year, toward the end of our time in New York, to almost 30,000 kilowatt-hours—and our house doesn't even have central air-conditioning. We bought a car shortly before we moved, and another one soon after we arrived, and a third one ten years later. (If you live in the country and don't have a second car, you can't retrieve your first car from the mechanic after it's been repaired. The third car was the product of a mild midlife crisis; it evolved into a necessity as soon as Laura and our son, John, became old enough to drive.) Ann and I both work at home, and therefore commute by climbing a flight of stairs, but, between us, we manage to drive more than 20,000 miles a year, mostly doing ordinary errands.[5] City dwellers who fantasize about living in the country usually picture themselves hiking, kayaking, gathering eggs from their own chickens, and engaging in other robust outdoor activities, but what you actually do when you move out of the city is move into a car, because public transit is nonexistent and most daily destinations are too widely separated to make walking or bicycling plausible as forms of transportation. Almost everything Ann and I do away from our house requires a car trip. The nearest movie theater is twenty minutes away, and so is the nearest large supermarket. Renting a DVD and later returning it consumes almost two gallons of gasoline, because

Blockbuster is ten miles away and each complete transaction involves two round trips. Quite often, we use a car when taking our dogs for a walk, so that the walk can begin somewhere other than our own yard. The office of our Manhattan pediatrician was in the lobby of our apartment building, an elevator ride away; the office of my Connecticut dentist is two towns over, a round trip of thirty-two miles. When we lived in New York, heat escaping from our apartment helped to heat the apartment above ours; nowadays, many of the BTUs produced by our very modern, extremely efficient oil-burning furnace leak through our two-hundred-year-old roof and into the dazzling star-filled winter sky above.

THE HISTORY OF CIVILIZATION IS A CHRONICLE OF DE-struction: people arrive, eat anything slow enough to catch, supplant indigenous flora with species bred for exploitation, burn whatever can be burned, and move on or spread out. No sensitive modern human can contemplate that history without a shudder. Everywhere we look, we see evidence of our recklessness, as well as signs that our destructive reach is growing. For someone standing on the North Rim of the Grand Canyon on a moonless night, the brightest feature of the sky is no longer the Milky Way but the glow of Las Vegas, 175 miles away.[6] Tap water in metropolitan Washington, D.C., has been found to contain trace amounts of caffeine, ibuprofen, naproxen sodium, two antibiotics, an anticonvulsive drug used to treat seizures and bipolar disorder, and the antibacterial compound triclocarban, which is an ingredient of household soaps and cleaning agents.[7] Modern interest in environmentalism is driven by a yearning to protect what we haven't ruined already, to conserve what we haven't used up, to restore as much as possible of what we've destroyed, and to devise ways of reconfiguring our lives so that civilization as we know it can be sustained through our children's lifetimes and beyond.

To the great majority of Americans who share these concerns, densely populated cities look like the end of the world. Because such places concentrate high levels of human activity, they seem to manifest nearly every distressing symptom of the headlong growth of civilization—the smoke, the filth, the crowds, the cars—and we therefore tend to think of them as environmental crisis zones. Calculated by the square foot, New York City generates more greenhouse gases, uses more energy, and produces more solid waste than any other American region of comparable size. On a map depicting negative environmental impacts in relation to surface area, therefore, Manhattan would look like an intense hot spot, surrounded, at varying distances, by belts of deepening green.

But this way of thinking obscures a profound environmental truth, because if you plotted the same negative impacts by resident or by household the color scheme would be reversed. New Yorkers, individually, drive, pollute, consume, and throw away much less than do the average residents of the surrounding suburbs, exurbs, small towns, and farms, because the tightly circumscribed space in which they live creates efficiencies and reduces the possibilities for reckless consumption. Most important, the city's unusually high concentration of population enables the majority of residents to live without automobiles—an unthinkable deprivation almost anywhere else in the United States, other than in a few comparably dense American urban cores, such as the central parts of San Francisco and Boston. The scarcity of parking spaces in New York, along with the frozen snarl of traffic on heavily traveled streets, makes car ownership an unbearable burden for most, while the compactness of development, the fertile mix of commercial and residential uses, and the availability of public transportation make automobile ownership all but unnecessary in most of the city. A pedestrian crossing Canal Street at rush hour can get the impression that New York is the home of every car ever built, but Manhattan actually has the lowest car-to-resident ratio of anyplace in America.

The apparent ecological innocuousness of widely dispersed populations—as in leafy suburbs or seemingly natural exurban areas, such as mine—is an illusion. My little town has about 4,000 residents, spread over 38.7 thickly wooded square miles (just eight fewer square miles than San Francisco), and there are many places within our town limits from which no sign of settlement is visible in any direction. But if you moved eight million people like us, along with our dwellings, possessions, vehicles, and current rates of energy use, water use, and waste production, into a space the size of New York City, our profligacy would be impossible to miss, because you'd have to stack our houses and cars and garages and lawn tractors and swimming pools and septic tanks higher than skyscrapers, and you wouldn't be able to build roads and gas stations fast enough to serve us, even if you could find places to put them. Conversely, if you made all eight million New Yorkers live at the density of my town, they would require a space equivalent to the land area of the six New England states plus Delaware and New Jersey.[8] Spreading people thinly across the countryside may make them feel greener, but it doesn't reduce the damage they do to the environment. In fact, it increases the damage, while also making the problems they cause harder to see and to address.

New York City is by no means the world's only or best example of the environmental benefits of concentrating human populations and mixing uses. Many large old cities in Europe—where the main population centers arose long before the automobile, and therefore evolved to be served by less environmentally disastrous means of getting around—are less wasteful than New York, and the most energy-efficient and least automobile-dependent cities in the world include a number of Asian ones, among them Hong Kong and Singapore. But New York is a useful example because it is familiar both to Americans and to people in the developing world, and because it proves that affluent people are capable of living comfortably while consuming energy and inflicting environmental damage at levels well below current U.S. averages. And—as is the case with all dense cities—New York's efficiencies are built-in and therefore don't depend on a total, sudden transformation of human nature. Even for people who live in sparsely populated areas far from urban centers, dense cities like New York offer important lessons about how to permanently reduce energy use, water consumption, carbon output, and many other environmental ills.

Thinking of crowded cities as environmental role models requires a certain willing suspension of disbelief, because most of us have been accustomed to viewing urban centers as ecological calamities. New York is one of the most thoroughly altered landscapes imaginable, an almost wholly artificial environment, in which the terrain's primeval contours have long since been obliterated and most of the parts that resemble nature (the trees on side streets, the rocks in Central Park) are essentially decorations. Quite obviously, this wasn't always the case. When Europeans first began to settle Manhattan, in the early seventeenth century, a broad salt marsh lay where the East Village does today, the area now occupied by Harlem was flanked by sylvan bluffs, and Murray Hill and Lenox Hill were hills. Streams ran everywhere, and beavers built dams near what is now Times Square. One early European visitor described Manhattan as "a land excellent and agreeable, full of noble forest trees and grape vines," and another called it a "terrestrial *Canaan*, where the Land floweth with milk and honey."[9]

But then, across a relatively brief span of decades, Manhattan's European occupiers leveled the forests, flattened the hills, filled the valleys, buried the streams, and superimposed an unyielding, two-dimensional grid of avenues and streets, leaving virtually no hint of what had been before. The earliest outposts of metropolitan civilization, such as it was, were confined to the island's southern tip, but in the eighteenth and nineteenth centuries settlement spread northward at an accelerating pace. In 2007, Eric Sanderson, a landscape ecologist who was completing a three-dimensional computer re-creation of precolonial Manhattan, told Nick Paumgarten, of *The New Yorker*, "It's hard to think of any place in the world with as heavy a footprint, in so short a time, as New York. It's probably the fastest, biggest land-coverage swing in history."[10] Picturing even a small part of that long-lost world requires a heroic act of the imagination—or, as in Sanderson's case, a vast database and complex computer-modeling software.

Given the totality of what has been erased, contemplation of New York's evolution into a megalopolis inspires mainly a sense of loss, and ecology-minded discussions of the city tend to have a forlorn air. Nikita Khrushchev, who visited New York in the fall of 1960, found the scarcity of foliage in the city depressing by comparison with Moscow, saying, "It is enough to make a stone sad."[11] In environmental triage, New York is usually consigned to the hopeless category, worthy of palliative care only. Environmentalists tend to focus on a handful of ways in which the city might be made to seem somewhat less oppressively man-made: by easing the intensity of development; by creating or enlarging open spaces around structures; by relieving traffic congestion and reducing the time that drivers spend aimlessly searching for parking spaces; by increasing the area devoted to parks, green-ery, and gardening; by incorporating vegetation into buildings themselves.

But such discussions miss the point, because in most cases changes like these would actually undermine the features that create the city's extraordinary efficiency and keep the ecological impact of its residents small. Spreading buildings out enlarges the distance between local destinations, thereby limiting the utility of walking and public transportation; making automobile traffic move more efficiently enhances the allure of owning cars and, inevitably, reduces ridership on the subway. Because urban density, in itself, is such a powerful generator of environmental benefits, the most critical environmental issues in dense urban cores tend to be seemingly unrelated matters like law enforce-ment and public education, because anxieties about crime and school quality are among the strongest forces motivating flight to the suburbs. By comparison, popular feel-good urban eco-projects like adding solar panels to the roofs of apartment build-ings are decidedly secondary, even irrelevant. Planting trees along city streets, always a popular initiative, has high environ-mental utility, but not for the reasons that people usually as-sume: trees are ecologically important in dense urban areas not because they provide temporary repositories for atmospheric carbon—the usual argument for planting more of them—but because their presence along sidewalks makes city dwellers more cheerful about dwelling in cities. Unfortunately, much conven-tional environmental activism has the opposite effect, since it reinforces the view that urban life is artificial and depraved, and makes city residents feel guilty about living where and how they do.

A dense urban area's greenest features—its low per-capita energy use, its high acceptance of public transit and walking, its small carbon footprint per resident—are not inexplicable anomalies. They are the direct consequences of the very urban characteristics that are the most likely to appall a sensitive friend of the earth. Yet those qualities are ones that the rest of us, no matter where we live, are going to have to find ways to emulate, as the world's various ongoing energy and environmental crises deepen and spread in the years ahead. In terms of sustainability, dense cities have far more to teach us than solar-powered mountainside cabins or quaint old New England towns.

THIS WAY OF THINKING SEEMS COUNTERINTUITIVE TO most Americans, including most environmentalists. Ben Jervey, in *The Big Green Apple*, a well-intentioned but frequently misleading guide to "eco-friendly living in New York City"—a concept that Jervey himself treats as oxymoronic—repeatedly misses the point about New York. "After growing up in a small town in Massachusetts," he writes in his preface, "I went off to pastoral Vermont to study and then work, all the while developing an appreciation and concern for the fragile state of the world's ecology. But as easy as it is to don a green hat up in Vermont, the beast that is New York City has the tendency to tear that noble lid off and throw it into a puddle of mud. Upon arriving in the big city I struggled to reconcile the environmentally concerned mind-set that comes so effortlessly in a place like Vermont with my new urban lifestyle. Of course sustainable living is easier in a Vermont township, where local produce is plentiful and every backyard is equipped with a compost bin."[12]

But this is exactly wrong. "Sustainable living" is actually much harder in small, far-flung places than it is in dense cities. Jervey cites New Yorkers' "overactive dependence" on fresh water as an example of their supposed wastefulness, and he marvels that the city's total use "amounts to well over one billion gallons per day."[13] A billion is a big number, to be sure, but New York City's population is more than thirteen times that of the entire state of Vermont, so the city's total consumption figures in any category will appear overwhelming in any direct comparison. It's per-capita consumption that is telling, though, and by that measure Vermonters use more water than New Yorkers do. They also use more than three and a half times as much gasoline—545 gallons per person per year versus 146 for all New York City residents and just 90 for Manhattan residents—with the result that, among the fifty states, pastoral Vermont ranks eleventh-highest in per-capita gasoline consumption while New York state, thanks entirely to New York City, ranks last. The average Vermonter also consumes more than four times as much electricity as the average New York City resident, has a larger carbon footprint, and generates more solid waste, backyard compost bins notwithstanding.[14]

Jervey is by no means alone. The prominent British environmentalist Herbert Girardet—who is an author, a documentary filmmaker, and a cofounder of the World Future Council—treats large cities mainly as environmental catastrophes. "The bulk of the world's energy consumption is *within* cities," he has written, "and much of the rest is used for producing and transporting goods and people *to and from* cities."[15] He proposes dramatically reducing urban energy consumption and making city dwellers less dependent on agricultural and other inputs from outlying areas, while improving overall energy efficiency through technological innovation. He has observed that cities cover just 3 or 4 percent of the earth's land area while accounting for 80 percent of the world's consumption of natural resources—as though population density were an ecological negative, and as though there were no meaningful distinction to be made between dense urban cores and lightly populated suburbs. Urban dwellers, by his way of thinking, are environmental freeloaders, parasitically drawing sustenance from the countryside, while people living at lower densities are more nearly at harmony with nature.[16] Girardet is a victim (and perpetuator) of the same optical illusion as Jervey.

New Yorkers themselves seldom fully appreciate the environmental virtues of their own way of living. On Earth Day 2007, the city announced an ambitious two-decade environmental initiative, called PlaNYC, which includes dozens of far-reaching proposals, among them the planting of more than a million trees, the collection of tolls from most private and commercial vehicles using the most traffic-clogged parts of Manhattan during the busiest times of the day, the imposition of a surcharge on the bills of the city's electrical customers, and other measures.[17] Actually implementing the plan has encountered the usual difficulties (shortly before Earth Day 2008, the state legislature killed the toll-collection scheme, which is known as "congestion pricing"), but one of the most striking features of the entire plan is how little recognition it gives to the numerous ways in which New York City's environmental performance is already exemplary, even extraordinary, at least in comparison with the rest of the United States. Shortly before the plan was made public, the mayor's office released a study showing that the city's buildings are responsible for 79 percent of its greenhouse-gas emissions—an ominous statistic, the study suggested, since the national average for buildings is just 32 percent. Daniel L. Doctoroff, a deputy mayor and the city official in charge of the plan, said, "We know we have to dramatically rethink the way we work with buildings"—probably an understatement, since the mayor's announced goal was to cut greenhouse-gas emissions by 30 percent by 2030.[18]

Cutting greenhouse-gas emissions is a fine idea, but in the case of the city's buildings the mayor's office obscured a far more important point. The proportion of emissions attributable to buildings in New York City is high because the number of cars, which are the main source of greenhouse emissions in the rest of the country, is extremely low in relation to the city's population: it's a sign of environmental success, not failure. Thinking in terms of proportions can only be misleading, since there's no way to decrease the percentage attributable to one element without increasing the percentage attributable to others: they're pieces of the same pie. Bringing down overall emissions levels is a worthy goal, but the mayor's emphasis was misplaced. The proportion of greenhouse-gas emissions attributable to buildings is higher in energy-efficient old European cities, too.

Equally misguided is the plan's proposal to add a surcharge to New Yorkers' electrical bills, since New York City residents, with an average of 4,696 kilowatt-hours per household per year, already consume less electricity than the residents of any other part of the country. (The average Dallas household, by contrast, uses 16,116 kilowatt-hours, more than three times as much.)[19] Many news reports about the study focused on the fact that New York City is responsible for almost 1 percent of all the greenhouse gases produced by the United States, and suggested that this share was shockingly huge—but they overlooked, or mentioned only in passing, the fact that the city contains 2.7 percent of the country's population, meaning that its carbon footprint is already remarkably low in comparison with that of other American communities. Mandating large reductions in categories in which New Yorkers already lead the nation is like trying to fight obesity by putting skinny people on diets.

Thinking of New York City's environmental record as something that might instruct and inspire others, rather than treating it as a candidate for emergency intervention, requires a major conceptual leap for many, even for those who deal directly with the city's relationship to the environment. In 2004, I called New York City's Department of Environmental Protection and told a member of that agency's staff that I was interested in talking to an expert about what I felt were ways in which New Yorkers are better environmental citizens than other Americans are. At first, she thought I was joking; later, I think, she decided I was nuts. "Why don't you call the Parks Department?" she said, finally, happy to be rid of me.

THE HOSTILITY OF MANY ENVIRONMENTALISTS TOWARD densely populated cities is a manifestation of a much broader phenomenon, a deep antipathy to urban life which has been close to the heart of American environmentalism since the beginning. Henry David Thoreau, who lived in a cabin in the woods near Concord, Massachusetts, between 1845 and 1847, established an image, still potent today, of the sensitive nature lover living simply, and in harmony with the environment, beyond the edge of civilization. Thoreau wasn't actually much of an outdoorsman, and his cabin was closer to the center of Concord than to any true wilderness, but for many Americans he remains the archetype—the natural philosopher guiltlessly living off the grid. John Muir, who was born twenty years after Thoreau and founded the Sierra Club in 1892, viewed city living as toxic to both body and soul.[20] The National Park Service, established by Congress in 1916, was conceived as an increasingly necessary corrective to urban life, and national parks were treated in large measure as sanctuaries from urban depravity. The modern environmental movement arose, in the 1960s and 1970s, when a growing sense of ecological crisis, first inspired nationwide by Rachel Carson's extraordinarily influential book *Silent Spring*,[21] combined with other social forces, including the civil rights movement, opposition to the Vietnam War, and the power of OPEC, to create a sense among large numbers of mainly young people that just about everything wrong with the United States was urban in essence, and could be combated only by establishing, or reestablishing, a direct connection to "the land." American environmentalists in every age have tended to agree with Thomas Jefferson, who, in 1803, dismissed "great cities" as "pestilential to the morals, the health and the liberties of man."[22]

Jefferson made that disparaging remark in a letter to Dr. Benjamin Rush, a fellow signer of the Declaration of Independence. Daniel Lazare, in *America's Undeclared War: What's Killing Our Cities and How We Can Stop It,* cites that letter as a key document in the history of what he identifies as an enduring national antagonism toward urban life. Recently, I asked Lazare whether he detected that same antagonism in the modern American environmental movement. "Unquestionably," he said. "Green ideology is a rural, agrarian ideology. It seeks to integrate man into nature in a very kind of direct, simplistic way—scattering people among the squirrels and the trees and the deer. To me, that seems mistaken, and it doesn't really understand the proper relationship between man and nature. Cities are much more efficient, economically, and also much more benign, environmentally, because when you concentrate human activities in confined spaces you reduce the human footprint, as it were. That is why the disruption of nature is much less in Manhattan than it is in the suburbs. The environmental movement is deeply stained with a sort of Malthusian current. It's anti-urban, anti-industrial, agrarian, primitivist. Manhattan seems to be a supremely unnatural place because of all the concrete and glass and steel, but the paradox is that it's actually more harmonious and more benign, in terms of nature, than ostensibly greener human environments, which depend on huge energy inputs, mainly in the form of fossil fuels. In order to surround ourselves with nature, we get in cars and drive long distances, and then build silly pseudo-green houses in the middle of the woods—which are actually extremely disruptive, and very, very wasteful."

To be sure, there has always been plenty to loathe about urban living. The history of large cities all over the world is a history of filth and squalor and disease. Benjamin Rush placed himself at tremendous personal risk in 1793, a decade before Jefferson's letter, while attempting to combat a yellow fever epidemic in Philadelphia, which was then both the nation's capital and, with a population of 55,000, its largest city. No one in those days knew how yellow fever was transmitted, but there was no local shortage of plausible explanations. The streets of Philadelphia, like the streets of most cities, were reeking, open sewers, and that particular summer the air had been made especially rank by the arrival from the Caribbean of a large shipment of spoiled coffee beans, which had been left to rot on the wharf and seemed to Rush to be the most likely cause of the disease.[23] Jefferson's letter made specific reference to that epidemic, which killed 4,000 Philadelphians (and caused Jefferson himself to flee the city, along with many other government officials and most of the city's wealthier inhabitants, including most of its physicians). "When great evils happen," Jefferson wrote to Rush, "I am in the habit of looking out for what good may arise from them as consolations to us, and Providence has in fact so established the order of things, as that most evils are the means of producing some good. The yellow fever will discourage the growth of great cities in our nation"—a providential result, in his view. He acknowledged that cities "nourish some of the elegant arts," but stated that "the useful ones can thrive elsewhere, and less perfection in the others, with more health, virtue & freedom, would be my choice."[24] New York City, he wrote twenty years later, "seems to be a Cloacina* of all the depravities of human nature."[25]

The early stirrings of industrialization magnified this sense of urban catastrophe. Human populations all over the world had always dumped their waste into the same lakes and streams from which they drew their drinking water, and the local consequences became more dire as the settlements grew, and as steady advances in human ingenuity outpaced awareness of the dangers

---

*Cloacina was the goddess of the Roman sewer system. The name comes from the Latin word for "sewer" or "drain."

posed by the effluents of prosperity. A source of drinking water for some early Manhattanites was Fresh Water Pond, also known as the Collect, a deep, seventy-acre spring-fed body of water just north of where Canal Street lies today. By 1800, though, the pond had become, according to various observers, "a shocking hole . . . foul with excrement, frog-spawn and reptiles," a "very sink and common sewer," and a heavily used dump for breweries, tanneries, and other toxin-generating commercial enterprises; within fifteen years it had to be filled in.[26] Throughout the city, the streets were mired in animal and human waste, and the air was thick with smoke and insects, and the shallow wells that provided drinking water for the city's residents were incubators of disease.

In 1832, cholera struck New York, killing 3,515, and its focus was the notorious neighborhood called Five Points, a foul slum that had arisen on the site of the filled-in Fresh Water Pond. (The same neighborhood provided the setting for Martin Scorsese's 2002 film *Gangs of New York*.) The epidemic inspired the same sort of panic and heroic but futile intervention that had characterized Philadelphia's response to yellow fever four decades earlier. A city newspaper reported, "The roads, in all directions, were lined with well-filled stage coaches, livery coaches, private vehicles and equestrians, all panic-stricken, fleeing the city, as we may suppose the inhabitants of Pompeii fled when the red lava showered down upon their houses."[27] It's no wonder that Jefferson felt, as he wrote to James Madison in 1787, "When we get piled upon one another in large cities as in Europe, we shall become corrupt as in Europe, and go to eating one another as they do there."[28]

Europeans viewed the same evolution with a similar sense of horror. In 1847, a Scottish visitor to England concisely summarized the dark side of that country's industrial progress, when he described the Irwell River as it flowed out of Manchester: "There are myriads of dirty things given it to wash, and whole waggon-loads of poisons from dye-houses and bleach-yards thrown into it to carry away; steam-boilers discharge into it their seething contents, and drains and sewers their fetid impurities; till at length it rolls on—here between tall dingy walls, there under precipices of red sandstone—considerably less a river than a flood of liquid manure, in which all life dies, whether animal or vegetable, and which resembles nothing in nature, except, perhaps, the stream thrown out in eruption by some mud-volcano."[29] The proposed solution was to reverse the direction of human migration—in effect, to create sprawl. In 1898, Ebenezer Howard, a British urban planner and the originator of the open-space-oriented development scheme known as the garden city movement, wrote, "It is wellnigh universally agreed by men of all parties, not only in England, but all over Europe and America and our colonies, that it is deeply to be deplored that people should continue to stream into the already over-crowded cities, and should thus further deplete the country districts." Howard, in support of this idea, quoted the cleric Frederic William Farrar, who had described large cities as "the graves of the physique of our race." Howard called the countryside "the symbol of God's love and care for man," and concluded that what Britain needed was "the spontaneous movement of the people from our crowded cities to the bosom of our kindly mother earth, at once the source of life, of happiness, of wealth, and of power."[30]

This idea—that city life is hopelessly demented and that the solution to urban problems is to spread out—has been with us ever since. It's the motivation for building suburbs, and it's still seductive; it's why I live where I live. But it's also a prescription for strip malls and expressways and tremendous waste, and it's the basis for the helter-skelter residential development which has turned out to be America's true manifest destiny. The Sierra Club has a national campaign called Challenge to Sprawl, the goal of which is to arrest the mindless conversion of undeveloped countryside into subdivisions and SUV-clogged expressways. But in a paradoxical way the Sierra Club itself has been a major contributor to sprawl, because the organization's anti-city ethos, which has been indivisible from its mission since the time of John Muir, has fueled the yearning for fresh air and elbow room which drives not only the preservation of wilderness areas but also the construction of disconnected residential developments and daily hundred-mile commutes. It also contributed to the popularity of SUVs and pickup trucks, both of which have been marketed by their manufacturers as "off-road" vehicles, designed to carry their nature-loving occupants into the great outdoors, even though just 6 percent of SUV owners ever actually operate their vehicles in four-wheel-drive mode. (Tom

McCarthy, in *Auto Mania*, points out that the names of SUVs almost always reinforce this wilderness fantasy: Blazer, Yukon, Pathfinder, Explorer, Expedition, Sierra. Among the advertising slogans for my car, a Subaru all-wheel-drive station wagon called an Outback, is "My other car is a pair of boots" and "It loves the outdoors as much as you do.")[31] Preaching the sanctity of open spaces helps to propel development into those very spaces, and the process is self-reinforcing because, as one environmentalist said to me, "Sprawl is created by people escaping sprawl." Wild landscapes are less often destroyed by people who despise wild landscapes than by people who love them, or think they do—by people who move to be near them, and then, when others follow, move again. Thoreau's cabin, a mile from his nearest neighbor, set the American pattern for creeping residential development, since anyone seeking to replicate his experience needed to move a mile farther along. Jefferson, too, embodied the ethos of suburbia. Indeed, he could be considered the prototype of the modern American suburbanite, since for most of his life he lived far outside the central city in a house that was much too big, and he was deeply enamored of high-tech gadgetry and of buying on impulse and on credit, and he embraced a self-perpetuating cycle of conspicuous consumption and recreational home improvement. The standard object of the modern American dream, the single-family home surrounded by grass, is a mini-Monticello.

Anti-urbanism still animates American environmentalism, and is evident in the technical term that is widely used for sprawl: "urbanization." This is unfortunate, because thinking of freeways and strip malls as "urban" phenomena obscures the ecologically monumental difference between Manhattan and Phoenix, or between Copenhagen and Kansas City, and fortifies the perception that population density is an environmental ill. In 2006, Melissa Holbrook Pierson, a writer who lives in a smallish town in the Hudson River Valley, in upstate New York, published a book called *The Place You Love Is Gone,* a deeply felt paean to the lost American landscape, the one obliterated by sprawl. At one point, driven by what she refers to as "lacerating nostalgia," she describes the nightmare transformation of Akron, Ohio, where she grew up in the late 1950s and early 1960s. "I can't help it if I want to live in the past!" she writes. "It's *my* past, the time forty years ago when there was still some wide-open space into which to insert some dreaming, and still some darkness at night over it." She even manages to weep a little over Hoboken, New Jersey, where she lived, mostly unhappily, as a young adult. Her bitterest emotions, though, she reserves for New York City, which she accuses of having destroyed a pastoral paradise in order to create the extensive upstate reservoir system that supplies its drinking water—of "rubbing its chin in contemplation of turning faraway valleys into pipes to service its water closets." The city's early-twentieth-century planners, anticipating the population growth to come, condemned farms and rural hamlets far from the city in order to build the extraordinary chain of reservoirs without which New York City could not exist, and Pierson describes this massive engineering project as "larceny." Her arguments persuaded Anthony Swofford, who reviewed the book in *The New York Times*. He wrote, "The story of New York City's water grab is astonishing, nearly unbelievable in its scope and greed," and he described the creation of the city's water system, as recounted by Pierson, as "rural slaughter for the survival of the city."[32]

But this is wrong. If New York City could somehow be dismantled and its residents dispersed across the state at the density of Pierson's current hometown, what remains today of pastoral New York state would vanish under a tide of asphalt. Dense urban concentrations of people, along with the freshwater reservoirs and other infrastructure necessary to support them, are not the enemies of the images she clings to. It is the existence of Manhattan, not the nostalgia of Baby Boomers, that makes the Catskills possible, and it's small-town residents, not subway-riding apartment dwellers, who foster strip malls.[33] You create open spaces not by spreading people out but by moving them closer together. Pierson does write, near the end of her book, that "it is the thousands of acres of uninhabited, forested land in the buffer zones of the New York City watershed that have preserved wilderness in the midst of an inexorably creeping urbanization."[34] But she doesn't acknowledge the role of her own form of nostalgia in the creation of the thing she hates. Many more acres of upstate pastoral paradise were destroyed by the steady spread of towns like hers than by the creation of the water supply system that makes it possible for New York City to exist. Building the city didn't fill the Hudson Valley with parking lots; fleeing the city did.

THE SIGNIFICANCE OF POPULATION DENSITY WAS ELU-cidated brilliantly in 1961 in a landmark book called *The Death and Life of Great American Cities*, by Jane Jacobs.[35] Jacobs up-ended many widely held ideas about how cities ought to be put together, and she has been celebrated ever since as an urban-planning iconoclast and visionary, but she could be viewed just as easily as a pioneering environmentalist. Indeed, Jacobs's book may be most valuable today as a guide to reducing the ecological damage caused by human beings, even though it scarcely men-tions the environment, other than by making a couple of passing references to smog.

The central idea of Jacobs's book is that density and diversity are the engines that make human communities work. She lived in Greenwich Village at the time,* and she had come to realize that the qualities she found most appealing about city life could be traced to the fact that she and her neighbors lived very near to one another, that their tightly spaced apartment buildings were of varying sizes and configurations, that residences were closely mixed with businesses, and that she and her neighbors

---

*She and her family moved to Toronto in 1968, primarily out of opposition to the Vietnam War. She died in 2006.

were not narrowly segregated by wealth. Society, she decided, has a critical mass. Spread people too thinly and sort them too finely, and they cease to interact; move them and their daily activities closer together, and the benefits cascade: their neighborhoods grow safer, they become more attuned to one another's needs, they have more restaurants and movie theaters and museums to choose from, and their lives, generally, become more varied and engaging. Jacobs's focus was on the vibrancy of city life, but the same urban qualities that she identified as enhancing human interaction also dramatically reduce energy consumption and waste. Placing people and their daily activities close together doesn't just make the people more interesting; it also makes them greener.

Unfortunately, her catalogue of the failures of modern urban planning also still applies, almost fifty years later, with little modification, all across America: "Low-income projects that become worse centers of delinquency, vandalism and general social hopelessness than the slums they were supposed to replace. Middle-income housing projects which are truly marvels of dullness and regimentation, sealed against any buoyancy or vitality of city life. Luxury housing projects that mitigate their inanity, or try to, with a vapid vulgarity. Cultural centers that are unable to support a good bookstore. Civic centers that are avoided by everyone but bums, who have fewer choices of loitering place than others. Commercial centers that are lackluster imitations of standardized suburban chain-store shopping. Promenades that go from no place to nowhere and have no promenaders. Expressways that eviscerate great cities."[36] These flaws, she argued persuasively, are not unavoidable; they are merely the products of our ongoing failure to understand what we really want.

Of course, living in densely populated urban centers still has many drawbacks, even though city streets, nowadays, are no longer ankle-deep in horse manure. New Yorkers at all income levels live in spaces that would seem cramped to Americans almost anywhere else. A friend of mine who grew up in a townhouse in Greenwich Village thought of his upbringing as privileged until, in prep school, he visited a classmate from the suburbs and was staggered by the house, the lawn, the cars, and the swimming pool, and thought, with despair, You mean I could live like this? Riding the subway can be depressing even to a committed transit supporter, and during the summer it is often distressingly dirty and hot. Ann's and my apartment was fourteen floors above Second Avenue, yet the noise from the street was so loud, even in the middle of the night, that we both slept with earplugs. Joggers in Manhattan have to weigh the benefits of exercise against the dangers of inhaling bus and taxi fumes while they run.

Density, for many of the same reasons that it makes people more efficient, makes disasters more efficient, too. On 9/11, the airplane that crashed in the Pennsylvania countryside killed the occupants of the plane, while the two planes that struck the World Trade Center killed thousands and could have killed tens of thousands if the circumstances had been slightly different. New York City's water supply enters the city through three tunnels, the loss of which would make the city uninhabitable. Epidemics, from the Black Death down, have inflicted their highest death tolls on dense urban populations, which, for the same reasons, are highly vulnerable to biological weapons. Rising sea levels won't be a direct problem in my little town, which is more than thirty miles from the coast, but even a small rise could cripple Manhattan's sewer system, which malfunctions during rainstorms even now. The most powerful earthquake known to have occurred in the continental United States—the so-called New Madrid quake, a series of four huge shocks that struck what is now southeastern Missouri in 1811 and 1812—rerouted the Mississippi River and could be felt on the East Coast, yet it killed fewer than a hundred people because the area above the epicenter was so sparsely settled. An earthquake of comparable magnitude occurring today along any known fault in Los Angeles or San Francisco would kill hundreds of thousands and create a public-health disaster beyond comprehension.

Nevertheless, barring a massive reduction in the earth's population, dense urban centers offer one of the few plausible remedies for some of the world's most discouraging environmental ills, including climate change. To borrow a term from the jargon of computer systems, dense cities are scalable, while sprawling suburbs and isolated straw-bale eco-redoubts are not. Anti-urban naturalists like Thoreau and Muir make poor guides for anyone struggling with the increasingly urgent problem of how to support billions of mobile, acquisitive, hungry human beings without triggering disasters that can't be contained. The environmental problem we face, at the current stage of our assault on the world's nonrenewable resources, is not how to make our teeming cities more like the countryside. The problem we face is how to make other settled places more like Manhattan, whose residents currently come closer than any other Americans to meeting environmental goals that all of us, eventually, will have to come to terms with.

NEW YORK'S EXAMPLE, ADMITTEDLY, IS DIFFICULT FOR others (or even New York itself) to imitate, because the city's remarkable population density is the result not of conscientious planning but of a succession of serendipitous historical accidents. The most important of those accidents was geographic: New York arose on a small island rather than on the mainland edge of a river or a bay, and the surrounding water served as a physical barrier to outward expansion. Manhattan is like a typical seaport turned inside out—a city with a harbor around it, rather than a harbor with a city along its edge. The deep water surrounding Manhattan and linking it to the ocean made the city easily accessible to large ships, and insularity gave the city more shoreline per square mile than other ports, major advantages in the days when one of the world's main commercial activities was moving cargoes between ships. (The sailing vessels lying at anchor along Manhattan's shoreline in that era were so numerous that they created, according to one description, "a circumferential forest of spars."[37]) Manhattan's physical isolation also drove early development inward and upward. The American cities with the next highest per-capita rates of transit use, San Francisco and Boston, are similarly constrained, since both are situated on island-like peninsulas, while the cities with the highest rates of automobile use—places like Atlanta, Phoenix, and Kansas City—are the ones that, throughout their history, have faced the fewest natural and political barriers to low-density horizontal expansion. The densest parts of Chicago are those abutting the western shore of Lake Michigan, which acted like a dam against the flux of population growth. Hong Kong is doubly insular, both geographically and geopolitically.

A second lucky accident was that Manhattan's street plan was created by merchants who were more interested in economic efficiency than in boulevards, parks, or empty spaces between buildings. In 1807, the state legislature appointed a local commission to "lay out streets, roads, public squares of such extent and direction as to them shall be most conducive to the public good," and the commissioners hired John Randel, Jr., a young surveyor, to create a detailed map of the island, most of which was still essentially wilderness. Randel and his assistants spent years meticulously measuring and documenting Manhattan's then complex topography—although on the plan he submitted to the commission, in 1811, the suggested street plan runs as it does now, in perfectly straight lines, forming a regular gridiron, as though the hills and streams did not exist. "The natural geography of the island was originally to be a factor in devising a street system," Robert T. Augustyn and Paul E. Cohen write in

Manhattan in Maps, "but there is little evidence in the . . . numbered parallel and perpendicular streets and avenues delineated on Randel's map that the topography of the island was even a consideration."[38] The plan adopted by the commissioners retained this feature, without which Manhattan's extreme density would have been harder to achieve. The plan also included only a handful of parks and public squares, all of them small. The commissioners' view regarding parks was that "vacant spaces" were made unnecessary by "those large arms of the sea which embrace Manhattan Island," thereby providing what they felt to be an adequate supply of fresh air and obviating the need to sacrifice developable real estate to recreation.[39] No one today would lay out such a large inhabited area with such a paucity of open space, but the relentlessness of the street plan is actually one of the keys to the city's continuing vitality—and to its greenness. One of Jane Jacobs's many arresting observations is that parks and other open spaces, if poorly planned, can actually make cities less livable, by creating dead ends that prevent people from moving freely between neighborhoods and by decreasing adjacent activity, a subject to which I'll return in chapter 4.[40] Manhattan's crush of architecture is paradoxically humanizing, because it brings the city's commercial, cultural, and other offerings closer together, thereby increasing their accessibility. It also makes the city greener, primarily by greatly reducing dependence on automobiles.

A third accident was that residential and commercial development were more thoroughly mixed in New York than they would later become in most other parts of the United States. The city, early in the twentieth century, was actually an originator and early adopter of zoning regulations—development rules intended to create sharp divisions between what, by then, had come to be viewed as incompatible human activities, by confining residential, commercial, and industrial uses in non-overlapping districts—but many parts of the city were already such a dense and fertile jumble as to be relatively impervious to the scheming of urban planners, a trait the city shared with the older cities of Europe. The liveliest and greenest parts of New York today are the ones that least conform to received American ideas about what should go next to what. In the rest of the country, zoning schemes that were conceived and implemented early in the twentieth century are among the most significant causes of sprawl, and among the most enduring impediments to public transit, since in many cases they make even moderate density impossible. In such municipalities, John Holtzclaw has written, "zoning requires front and side yard setbacks, wide streets and two or more off-street parking places, reducing densities and separating destinations. Many suburbs prohibit sidewalks and convenient nearby markets, restaurants, and other commerce. These government mandates force destinations farther apart, lengthening trips, such that nonautomotive modes become less viable."[41]

A fourth accident was the fact that by the early 1900s most of Manhattan's lines had been filled in to the point where not even Robert Moses, the metropolitan area's "master builder," could redraw them to accommodate the automobile.[42] Before cars, people lived close to other people to survive; with cars, proximity became less important—indeed, it became undesirable. Henry Ford, who viewed urban life with as much distaste as Jefferson had, called the city a "pestiferous growth" and thought of his cars as tools for liberating humanity. In 1932, John Nolen, a prominent Harvard-educated urban planner and landscape architect who embraced Ford's notion of urban liberation-by-automobile, said, "The future city will be spread out, it will be regional, it will be the natural product of the automobile, the good road, electricity, the telephone, and the radio, combined with the growing desire to live a more natural, biological life under pleasanter and more natural conditions."[43] This is the very formula for sprawl, and most of the country has followed it.

New York City's apparent urban antithesis, in terms of automobile use, is metropolitan Los Angeles, whose metastatic outward growth has been virtually unimpeded by the lay of the land, whose early settlers came to the area partly out of a desire to create space between themselves and others, and whose main development began late enough to be shaped mainly by the needs of cars. But a more telling counterexample is Washington, D.C., whose basic layout was conceived at roughly the same time as Manhattan's. The District of Columbia's original plan was created by an eccentric French-born engineer and architect named Pierre-Charles L'Enfant, who befriended General Washington during the Revolutionary War and asked to be allowed to design the capital. L'Enfant was notoriously hard to get along with, and he was fired after little more than a year, in 1792, but many of modern Washington's most striking features are his: the broad, radial avenues; the hublike traffic circles; the sweeping public lawns and ceremonial spaces.

Washington is commonly viewed as the most intelligently beautiful—the most European—of large American cities, and it is, indeed, a city of restrained proportions and stirring metropolitan vistas. Ecologically, though, it's a mess and not truly European at all. The city was designed in part to make true density impossible; and because the federal government grew more slowly than the national economy, there was no pressure to abandon that early ideal. L'Enfant's expansive avenues were easily adapted to automobiles, and the low, widely separated buildings (whose height is limited by law) stretched the distance between destinations: keeping civilization low makes it wide. There are many pleasant places in Washington to go for a walk, but it is actually difficult to get around the city on foot. The wide avenues are hard to cross, the huge traffic circles are like obstacle courses, and the grandiloquent empty spaces thwart pedestrians. Many parts of Washington, furthermore, are relentlessly homogeneous. Dignified public buildings abound on Constitution Avenue, but good luck finding a dry cleaner, a Chinese restaurant, or a grocery store. The D.C. subway system is modern, clean, and extensive, but no one with a car feels compelled to take the train because there's always a place to park. The city's horizontal, airy design has also pushed development far into the surrounding countryside. One of the fastest-growing counties in the United States is Loudoun County, Virginia, at the rapidly receding western edge of the Washington metropolitan area. When cities are built on a "human" scale, they virtually force the creation of vast suburbs, with miles of freeways, long commutes, traffic jams, and shopping malls. The District of Columbia was a thing of beauty when the region surrounding it was relatively empty of human beings, but the city, as governed by its own design and land-use rules, is structurally unable to absorb its own growth. The sprawl of metropolitan Washington is not a perversion of L'Enfant's plan; it's the logical result.

ONE OF THE MOST ABUSED WORDS IN THE ENGLISH language in recent years, without a doubt, has been "sustainable." Like "solution"—a vaporous buzzword ubiquitous in corporate slogans—it signifies both anything and nothing. Hundred-thousand-dollar kitchen renovations are described as sustainable if the doors of the new cabinets are veneered with bamboo; concept cars are called sustainable if their seats are made with soy-based foam. A similar fog of meaninglessness characterizes almost any recent marketing effort with an environmental theme. An article in *The New York Times* in 2007 provided a humorous catalogue of contradictions from the shelves of Home Depot, which was running a green promotion it called Eco Options: "Plastic-handled paint brushes were touted as nature-friendly because they were not made of wood. Wood-handled paint brushes were promoted as better for the planet because they were not made of plastic."[44] In 2008, Discovery Communications launched Planet Green, an environmentally oriented cable-television channel, whose "exclusive automobile sponsor" was General Motors.[45] The cover story of the March–April 2008 issue of *Correctional News*, a magazine for people who run prisons, was "Greening the Big House: Sustainability in Corrections."[46] If you write to the makers of Annie's macaroni and cheese (a family favorite), they'll send you a "BE GREEN, Help the Earth Live" bumper sticker for your car, to let others know "that you're an Earth advocate, and that you care about what happens to our wondrous blue and green planet."[47]

Most of the products, technologies, and practices popularly touted as sustainable are not sustainable at all. Driving a gas-electric hybrid automobile is more environmentally benign, mile for mile, than driving a Hummer, but hybrids are not sustainable, because they require petroleum and the world's supply of petroleum is finite. Buying locally grown food can put interesting, wholesome meals on people's dinner tables, but "locavorism" is not sustainable as a strategy for feeding the world, or even northwestern Connecticut, because spreading populations across arable regions at densities low enough to make agricultural self-sufficiency feasible would be an environmental and economic disaster. A private mini-hydroelectric plant powered by a rushing stream may enable its owner to disconnect from the public power grid, but such power plants are not sustainable for anyone but their owners, because the earth's population could not survive in any arrangement of dwellings which would enable every residence to generate its own electricity. In the very long run, of course, life itself is unsustainable, no matter what we human beings do or fail to do, because the sun will eventually burn out. Over time spans shorter than eons, though, uncertainties abound. The way we Americans live now is clearly unsustainable, since we are rapidly depleting the natural resources on which we've built our turbocharged way of life. The cherished secret hope of most of us—that some sudden technological breakthrough will enable our children and grandchildren to live the way we live now, except with smaller cars and larger recycling bins—is patently a fantasy, at least until the physicists get nuclear fusion sorted out.

The crucial fact about sustainability is that it is not a micro phenomenon: there can be no such thing as a "sustainable" house, office building, or household appliance, for the same reason that there can be no such thing as a one-person democracy or a single-company economy. Every house, office building, and appliance, no matter where its power comes from or how many of its parts were made from soybeans, is just a single small element in a civilization-wide network of deeply interdependent relationships, and it's the network, not the individual constituents, on which our future depends. Sustainability is a context, not a gadget or a technology. This is the reason that dense cities set such a critical example: they prove that it's possible to arrange large human populations in ways that are inherently less wasteful and destructive.

In 1997, in Kyoto, Japan, representatives of most of the world's countries, after two and a half years of sometimes highly contentious negotiations, adopted a protocol intended to reduce global production of greenhouse gases. The United States signed the original agreement but pulled out in 2001, becoming one of only two of the original signatories to refuse to ratify the plan, which went into effect in 2005. (The other holdout was Kazakhstan.) America's intransigence has infuriated many environmentalists, at home and elsewhere, but in practical terms the impact of our refusal to sign has been zero. So far, the most effective way for a country to cut its carbon output has been to suffer a well-timed industrial implosion, as Russia did after the collapse of the Soviet Union, in 1991. The Kyoto benchmark year is 1990, when the smokestacks of the Soviet military-industrial complex were still blackening the skies. By the time Vladimir Putin ratified the protocol, in 2004, Russia was already certain to meet its goal for 2012. The countries with the best emissions-reduction records—Ukraine, Latvia, Estonia, Lithuania, Bulgaria, Romania, Hungary, Slovakia, Poland, and the Czech Republic—were all parts of the Soviet empire and therefore look good for the same reason. Ted Nordhaus and Michael Shellenberger, in their 2007 book *Break Through: From the Death of Environmentalism to the Politics of Possibility*, write, "Germany and Britain have reduced their emissions, but most of those reductions were due to the collapse of the British coal-mining industry in the 1980s and the collapse of East German heavy industry and power generation after the reunification of

Germany. Little of the reduction in Britain or Germany is attributable to regulatory actions taken by the European Union or national governments in the effort to reduce greenhouse-gas emissions. Greenhouse-gas emissions throughout the rest of Europe and the rest of the developed world have either remained steady or increased."[48]

Canada ratified the Kyoto Protocol, and its experience is suggestive because its economy and per-capita oil consumption are similar to those of the United States. Canada's Kyoto target is a 6 percent reduction from 1990 levels. By 2006, however, despite the expenditure of billions of dollars on climate initiatives, its greenhouse-gas output had increased to 122 percent of the goal. And Canada's post-Kyoto record looks even worse if you include LULUCF (land use, land-use change, and forestry), a calculation intended to reflect the greenhouse impact of timber harvesting, land clearing, and similar activities; including LULUCF, the increase in Canada's emissions was more than twice as high. In 2006, Canada's environmental minister described his country's Kyoto targets as "impossible."[49]

The explanation for Canada's difficulties isn't complicated: the world's principal source of man-made greenhouse gases has always been prosperity. That relationship is easy to see, now that the global recession has flipped it onto its back: shuttered factories don't spew carbon dioxide; the unemployed drive fewer miles and turn down their furnaces, air conditioners, and swimming-pool heaters; struggling corporations and families cut back on air travel; even affluent people buy less throwaway junk. Gasoline consumption in the United States fell almost 6 percent in 2008. That was the result not of a sudden greening of the American consciousness but of the rapid rise in the price of oil during the first half of the year, followed by the full efflorescence of the current economic mess.

What would it take, short of utter economic collapse, for a prosperous First World population to reduce its carbon output and other environmental impacts permanently? The standard prescription is familiar: less reliance on fossil fuels, more reliance on renewable energy (and uranium), increased efficiency, reduced waste, more buses, fewer incandescent lightbulbs, more recycling. These and other elements, to be sure, will become increasingly important parts of our lives with every month that passes, but decades of experience have shown that the measurable results of our conscious efforts to use less are seldom as significant as forecast, and that reductions in waste are typically offset or exceeded by increases in consumption.

These discouraging realities make urban density even more significant as an environmental tool. Cutting back overall U.S. per-capita greenhouse emissions to New York City's current level would require a national reduction of 71 percent—a feat that not even the wildest Kyoto optimist thinks is remotely achievable. Yet New York's record is not the result of a massive, expensive environmental campaign; it's the result of New Yorkers living the way New Yorkers have always lived. The city's efficiencies, like the efficiencies of all dense urban cores, are built into the fabric of the place, and they don't depend on an unprecedented commitment to sacrifice and compliance by environmentally concerned citizens. In fact, New Yorkers themselves, when informed that their per-capita energy consumption is the lowest in the United States, usually express surprise. They don't generate less carbon because they go around snapping off lights.

Granted, directly comparing New York's greenhouse emissions with those of the rest of the country is unfair to much of the rest of the country, because the city couldn't exist without massive agricultural, industrial, and other inputs from far beyond its borders, and is therefore responsible for emissions occurring elsewhere. But all other American communities are subject to this same interdependence, and, even if they weren't, New York's example would still be significant because the city proves that tremendous environmental gains can be achieved by arranging infrastructure in ways that make beneficial outcomes inescapable and that don't depend on radically reforming human nature or implementing technologies that are currently beyond our capabilities or our willingness to pay. At an environmental presentation in 2008, I sat next to an investment banker who was initially skeptical when I explained that Manhattanites have a significantly lower environmental impact than other Americans. "But that's just because they're all crammed together," he said. Just so. He then disparaged New Yorkers' energy efficiency as "unconscious," as though intention trumped results. But unconscious efficiencies are the most desirable ones, because they require neither enforcement nor a personal commitment to cutting back. I spoke with one energy expert, who, when I asked him to explain why per-capita energy consumption was so much lower in Europe than in the United States, said, "It's not a secret, and it's not the result of some miraculous technological break-through. It's because Europeans are more likely to live in dense cities and less likely to own cars." In European cities, as in Manhattan, in other words, the most important efficiencies are built-in. And for the same reasons.

This is not necessarily a message that Americans like to hear, or that environmentalists like to give. The Sierra Club's website features a slide-show-like demonstration that illustrates how various sprawling suburban intersections could be transformed into far more appealing and energy-efficient developments by imple-menting a few modifications, among them widening the side-walks and narrowing the streets, mixing residential and commercial uses, moving buildings closer together and closer to the edges of sidewalks (to make them more accessible to pedestrians and to increase local density), and adding public transportation—all fundamental elements of the widely discussed anti-sprawl strategy known as Smart Growth. In a 2004 telephone conversation with a Sierra Club representative involved in Challenge to Sprawl, I said that the organization's anti-sprawl suggestions and the modified streetscapes in the slide show shared many significant features with Manhattan—whose most salient characteristics include wide sidewalks, narrow streets, mixed uses, densely packed buildings, and an extensive network of subways and buses. The representative hesitated, then said that I was essentially correct, although he would prefer that the program not be described in such terms, since emulating New York City would not be considered an appealing goal by most of the people whom the Sierra Club is trying to persuade. The truth, though, is inescapable. In a world of nearly 7 billion people and counting, sustainability, if it can be achieved, will look a lot more like midtown Manhattan than like rural Vermont.

The environmental lessons that New York City offers are not necessarily easy to apply—and, even to New Yorkers, they can often be difficult to discern—but the most important of them can be summarized simply:

- *Live smaller:* The average American single-family house doubled in size in the second half of the twentieth century, and the size of the average American household shrunk. Oversized, under-occupied dwellings permanently raise the world's demand for energy, and they encourage careless consumption of all kinds. In the long run, big, empty houses are no more sustainable than SUVs or private jets, no matter how many photovoltaic panels they have on their roofs. As the cost of energy inevitably rises in the years ahead, and as the long-term environmental and economic consequences of our accustomed levels of wastefulness become clearer and more dire, we are going to need to find ways to reduce the size of the spaces we inhabit, heat, cool, furnish, and maintain. (A notable countertrend: while the typical American single-family house was doubling in size, rising real estate values in New York City were reducing the size of the living space of the average Manhattan resident, thereby making it more efficient.)

- *Live closer:* The main key to lowering energy consumption and shrinking the carbon footprint of modern civilization is to contract the distances between the places where people live, work, shop, and play. Unfortunately, the steady enlargement of the American house was accompanied by the explosive growth of low-density subdivisions and satellite communities linked by networks of new highways and inhabited by long-distance commuters. Living closer to one's daily destinations, Manhattan-style, reduces vehicle miles traveled, makes transit and walking feasible as forms of transportation, increases the efficiency of energy production and consumption, limits the need to build superfluous infrastructure, and cuts the demand for such environmentally doomed extravagances as riding lawnmowers and household irrigation systems. The world, not just the United States, needs to pursue land-use strategies that promote high-density, mixed-use urban development, rather than sprawl.

- *Drive less:* Making automobiles more fuel-efficient isn't necessarily a bad idea, but it won't solve the world's energy and environmental dilemmas. The real problem with cars is not that they don't get enough miles to the gallon; it's that they make it too easy for people to spread out, encouraging forms of development that are inherently wasteful and damaging. Most so-called environmental initiatives concerning automobiles are actually counterproductive, because their effect is to make driving less expensive (by reducing the need for fuel) and to make car travel more agreeable (by eliminating congestion). What we really need, from the point of view of both energy conservation and environmental protection, is to make driving costlier and less pleasant. And that's as true for cars that are powered by recycled cooking oil as it is for cars that are powered by gasoline. In terms of the automobile's true environmental impact, fuel gauges are less important than odometers. In the long run, miles matter more than miles per gallon. As we make *cars* more efficient, we must compensate by making *driving* less so—a goal both harder to attain and less likely to be embraced by drivers themselves.

None of these imperatives will be easy to implement. But New York and the world's other dense cities point the way. Those cities' long-term value as role models has yet to be widely embraced, partly because many of the benefits of urban density are counterintuitive, and partly because most Americans, including most environmentalists, are more likely to think of places like Manhattan as exasperating environmental problems than as tantalizing sources of environmental solutions. New York is the place that's fun to visit but you wouldn't want to live there. What could it possibly teach anyone about being green?

Two

Liquid Civilization

Every serious discussion of the environment—every book, every documentary, every television news report, every magazine article, every lecture, every dire warning—is ultimately about oil, whether it specifically mentions oil or not. All the exasperatingly difficult environmental challenges we face today, large and small, are consequences of the explosive growth, during the past century or so, of the complex apparatus of modern civilization, and that growth has been engendered and nurtured and driven and amplified by oil, without which it could not have occurred. Most of the major environmental problems we currently face are the result of oil's prodigious abundance during the twentieth century; most of the problems we will face going forward will be the result of oil's increasing scarcity and cost during the twenty-first.

The cost of fossil fuels is an important part of the cost of everything we buy and everything we do. Virtually all I own was created on machines that were powered by oil, coal, or natural gas, or was delivered to me over a transportation network fueled mainly by oil, or was made from components that were synthesized from oil, or was grown with the help of chemicals manufactured from oil and natural gas. Most of the clothes I'm wearing contain materials that were made from oil derivatives, and the rest were made from plant or animal fibers that couldn't have been produced or gathered or woven or packaged or shipped without fossil fuels. In 2007, the man who mows my lawn added a 4 percent fuel surcharge to his bills. Another of his customers felt that there was something underhanded about that, and complained (despite the obvious, direct connection between the cost of oil and the cost of operating machines powered by gasoline), but the truth is that every product and service we buy includes, and has always included, a fuel surcharge, because the cost of doing any business necessarily reflects the cost of the fuel that is consumed in the course of doing it. For a century or so, that implicit surcharge was so small as to be invisible to most of us; now it has the potential to reconfigure the economies of the world. As the cost of fossil fuels goes up and down, the size of the fuel surcharge changes, too, on everything we buy and do, whether the surcharge is itemized on the bill or not.

As global oil prices rose from around ten dollars a barrel in 1999 to almost fifteen times that amount in the summer of 2008, Americans tended to think of the increase solely in terms of its likely impact "at the pump," on the cost of driving their own car. Inhabitants of my part of the country, where many houses are heated with oil, had a somewhat broader sense of the significance, as did people who work in industries that are directly and powerfully affected by the price of oil, such as airlines. For the most part, though, Americans had—and still have—a narrow understanding of the implications of expensive oil, equating it mainly, if not entirely, with expensive gasoline. Even there, most of us are susceptible to the well-documented inability of human beings to make rational economic calculations, and we are highly likely to either grossly overestimate or grossly underestimate the specific consequences of rising or falling fuel costs. In the summer of 2004, when oil reached forty dollars a barrel and gasoline was selling for a little more than two dollars a gallon, *The New York Times* ran a front-page story titled "Many Feeling Pinch After Newest Surge in U.S. Fuel Prices." The article mentioned a Meals on Wheels program in a Denver suburb which had lost two volunteer drivers because of the price of gasoline and was in danger of losing more, including a married couple whom the reporter had interviewed. That couple had driven a thirty-mile delivery route once a week for fifteen years but had now begun to worry that gas prices would soon force them to quit because, as the wife explained, "you have to give up something to be able to do something else."[1] Yet the increase in total fuel expense which had prompted her anxieties (if you added up the numbers deducible from the article) worked out to less than seventy-five cents a week—hardly the sort of financial penalty that causes sensible people to undertake significant lifestyle changes. Most of us, after all, don't go out of our way to avoid far greater financial penalties, such as three-dollar ATM transaction fees and the decline in fuel efficiency which comes from exceeding highway speed limits, accelerating rapidly, braking suddenly, or driving on underinflated tires.

Conversely, in 2008, a neighbor of mine sent two dozen friends and relatives an already much-forwarded e-mail, which he described in his covering note as "one of the few things that I have read lately in this area that make sense." Its thesis was that high gasoline prices are the result of corporate greed and can be reversed through stubborn consumer resistance. "This was sent by a retired Coca Cola executive," the forwarded e-mail said. "It came from one of his engineer buddies who retired from Halliburton. If you are tired of the gas prices going up AND they will continue to rise this summer, take time to read this please." The recommended course of action was simple: "For the rest of this year, DON'T purchase ANY gasoline from the two biggest companies (which are now one), EXXON and MOBIL. If they are not selling any gas, they will be inclined to reduce their prices. If they reduce their prices, the other companies will have to follow suit." In order for this plan to be fully effective, the e-mail's anonymous author wrote, 300 million participants would have to be enlisted—easily accomplished if each recipient would forward the message to ten friends. "If you don't understand how we can reach 300 million and all you have to do is send this to 10 people. . . . Well, let's face it, you just aren't a mathematician. But I am, so trust me on this one." What could be simpler? A little viral spamming and the world's energy problems go away.

Reactions like these, in both directions, are the products not just of ignorance but of powerlessness. You feel that you are being swept along by forces beyond your influence, and you urgently want to do something—but what? Later, usually, the sense of crisis gradually dissipates, and old habits reassert themselves. (My spamming neighbor and his wife, despite their concern about gas prices, continued to drive a full-sized Cadillac and a jumbo SUV.) One of our best and worst traits, as a species, is our often truly remarkable ability to absorb bad news and go on. Between 1987 and 2007, as scientists were becoming more certain and more specific about the consequences of our extraordinarily robust dependence on oil, new American cars actually grew 29 percent in weight (from 3,220 to 4,150 pounds) and 89 percent in horsepower (from 118 to well over 200), while declining 2 percent in fuel efficiency, and U.S. gasoline consumption increased by about a third. Sales of oversized cars have fallen dramatically since early 2008, and the automobile industry has been transformed to the verge of financial collapse. But people still buy cars that are too big, and people who have switched to hybrids don't necessarily drive them in ways that maximize mileage. If you want to make a typical American driver even angrier than he is about high gasoline prices, get in front of him on a two-lane highway and drive the speed limit.

The year 2008 provided an almost surreal illustration of the complex network of relationships between fossil fuels, the environment, and economics. During the first half of that year, oil's price rose rapidly and fluctuated for weeks at about $150 a barrel, an all-time high. Those prices proved to be transitory—within a few months the cost of oil had fallen again by more than a hundred dollars a barrel—but while they lasted they allowed the world to experience a sort of dress rehearsal for the energy crisis we must still endure, at some point in the coming months or years, once the world's demand for oil has permanently outstripped the world's ability to supply it.

When the price of a gallon of gasoline first approached three dollars, I read a news story in which a resident of a northeastern suburb was quoted as saying that driving was an irreducible element of his life and that he would cut back his spending in any and all other areas to the degree necessary to cover increases in the cost of operating his car. As that wavering Meals on Wheels driver in Denver said, "You have to give up something to be able to do something else." This is what we all do; it's the ordinary, everyday juggling of financial priorities. Yet one person's forgone expenditure is another person's ruined livelihood. That pinched suburban driver, in order to keep his gas tank full, cuts back on restaurant meals, and then the suddenly struggling restaurant owner decides to go another year without repainting his peeling building, and then the now idle housepainter decides that he can no longer afford to trade in his sputtering pickup truck, and on and on. Meanwhile, the rising cost of oil attacks all of these same people along a second front, by pushing up the cost of food and electricity and paint and steel and all the other raw materials of their and everyone else's occupation. Because the cost of energy is blended into the cost of everything, changes in the cost of energy have the power to transform lives. This same effect operated in the opposite direction beginning in late 2008, when the plunging cost of oil softened the impact, for American consumers, of the spreading global recession.

Rising oil prices after 1999 were also responsible for America's misguided decision to promote the production of ethanol as a gasoline substitute and extender. Ethanol has been viewed as the motor fuel of the future for more than a century—Henry Ford, anticipating eventual petroleum shortages, designed the Model T to run also on alcohol—but it has many disadvantages, both economically and environmentally, and it is not the energy panacea it is often presented to be. U.S. ethanol production is still minuscule, relatively speaking. In 2006, it amounted to less than 5.5 billion gallons. Because alcohol, when burned, yields only about two-thirds as much energy as gasoline—a fact that explains why your car gets better mileage on pure gasoline than it does on any ethanol blend—those 5.5 billion gallons provided the energy equivalent of 3.5 billion gallons of regular unleaded, or about enough to keep all of America's gasoline-powered engines running for something like two weeks. Yet producing even that modest amount required 20 percent of the U.S. corn crop that year, along with billions of dollars' worth of ill-considered federal subsidies and import restrictions, and contributed to higher prices at American gas stations and grocery stores.* It also boosted the price of natural gas—corn cultivation depends heavily on nitrogen fertilizers, which are manufactured primarily from natural gas, at the rate of approximately 33,000 cubic feet of gas per ton of fertilizer, accounting for approximately 5 percent of the world's annual gas production[2]—and exacerbated food shortages all over the world.[3] Global food prices rose 83 percent between 2005 and 2008, mainly because of increases in direct energy and fertilizer costs but also partly because of the diversion of foodstuffs, in the United States and elsewhere, into the production of biofuels. (The price of corn alone rose 124 percent between early 2006 and early 2008, from $250 a metric ton to $560.)[4] The impact has been especially high in the Third World, where even small increases in the price of food can be devastating to entire populations. In 2008, a survey by the World Bank found that many countries had acted to prevent exports of indigenous foodstuffs, in the hope of maintaining local supplies

---

*In 2008, the share of the U.S. corn crop devoted to ethanol production was expected to rise to a third. Not all of a corn kernel is consumed when ethanol is made from it, and the parts that are left have other economic uses, including as animal feed, but using ethanol as a motor fuel still creates more environmental problems than it solves. Ethanol made from sugarcane—the kind of ethanol produced in Brazil, a country that is often offered as an energy role model for the United States—is much more cost- and energy-efficient than the corn-based variety, but the Brazilian sugarcane industry depends on massive amounts of dangerous, grueling, low-paying manual labor, of a type that has been characterized as modern slavery, and it poses the same agricultural-resource dilemma that corn-based ethanol does. You can't grow both fuel and food on the same acre of land, and global demand for ethanol increases the economic pressure to destroy forests in order to create new cropland.

and holding down local prices—the food version of the isolationist national energy policy promoted by many U.S. politicians. These efforts, *The Wall Street Journal* reported, actually had the opposite effect, by distorting world food markets and driving prices higher. (The article was accompanied by a photograph of a food riot in Haiti.)[5] Limiting exports—global "locavorism"—necessarily creates shortages in importing countries, as well as inviting trade retaliation. Susan Schwab, who was the U.S. Trade Representative between 2006 and 2009, has said, "If every country in the world decided it wanted to produce its own food for consumption, there would be less food in the world, and more people would be hungry."[6] In the United States, agriculture is one of the few reliable export industries. Restricting American-grown foodstuffs to consumption by Americans would drive up food prices at home, exacerbate food shortages abroad, and eliminate one of the few available tools for reducing the staggering U.S. trade deficit.

Oil's high price in the summer of 2008 had several salutary impacts: it stimulated interest and investment in public transportation and renewable energy technologies; it led to a drop in many forms of wasteful consumption, including frivolous air travel; and it radically and permanently transformed the decades-old codependent relationship between automobile manufacturers and the drivers of laughably oversized cars and trucks. But because the cost of energy affects the cost of nearly everything, oil's price rise also had a broad range of negative consequences, in addition to the obvious ones. The ongoing global credit crisis arose from complex causes stretching back many years; but it had short-term triggers, too, and the steadily rising price of oil was one of them. As the price of fuel contributed to business closures, job layoffs, and rapid increases in the cost of food, clothing, medical care, and travel of all kinds, American home-owners at the margin were pushed beyond their ability to adjust—especially if the houses they had stretched to buy were in the newest, most distant suburbs, whose residents face the longest, most expensive commutes. ("Drive until you qualify" is the mortgage broker's expression of the inverse relationship between fuel consumption and what buyers perceive to be affordable real estate.) A new house that was barely within reach when oil was $20 a barrel became a financial land mine when oil was $145 a barrel. The global credit bubble would have burst regardless—for too many years, too many financial institutions had aggressively lent money they didn't have to people who couldn't pay it back, making it easier for all of us to buy things we couldn't afford—but the cost of energy was among the proximate causes, and it will always be one of the main factors determining the health of any of the world's economies, from the poorest to the richest.

In much the same way that oil's record prices in the summer of 2008 constituted an unsettling dress rehearsal for the world's permanent fossil-fuel crisis, the worst parts of which are still to come, the spreading global credit implosion has provided a preview of another kind, by indirectly demonstrating how difficult it will be for the world to achieve meaningful reductions in fuel consumption and carbon output through the kinds of conscientious political action that are anticipated by well-meaning agreements like the Kyoto Protocol. As the world's economies began to seize up in 2008, their carbon footprints shrank, too. What may seem less obvious is that the same relationship must operate in the other direction, as well: scaling back civilization's most environmentally destructive processes and activities also means scaling back what we think of as prosperity, because prosperity as we know it is founded on the ready availability of inexpensive fuels. Cutting back on frivolous air travel reduces fuel consumption and carbon output, but it also weakens economies, by undercutting businesses that depend for their existence on the spending of air travelers. There has been much talk of reviving the economy with green jobs and green industries, but that won't be as easy as it may sound. Creating green jobs is different from creating new jobs, since green jobs, if they're truly green, displace non-green jobs— wind-turbine mechanics instead of oil-rig roughnecks—probably a zero-sum game, as far as employment is concerned. The biggest effort we've made, up to now, to create a green industry—the billions we've wasted on ethanol—has put people to work and enriched some of the participants, but, overall, it's been an environmental and economic disaster, and not just for us.

A further consequence is that the paradoxical environmental benefits of economic recession, though real, are tremendously fragile, because they are vulnerable to intervention by governments that, understandably, would like to put people back to work and get them consuming again—for example, through programs intended to revive consumer spending (which has a big carbon footprint), and through public-investment projects aimed at doing things like building new roads and airports (ditto). In my area, stimulus money has already been allocated for two very non-green projects: the construction of new roads to serve two planned suburban industrial parks. The point is not that recessions are good—although, from an environmental perspective, there are things to be said for them—but that people's best intentions regarding conservation and carbon reduction inevitably run into the human reality of foreclosure and bankruptcy and unemployment. How do we persuade people to drive less—an environmental necessity—while also encouraging them to revive our staggering economy by filling their garages with new cars?

The world's economic and energy crises are linked, and they are fundamentally similar, because credit and fossil fuels are both forms of leverage. Oil, coal, and natural gas are multipliers of human labor in very much the same way that credit is a multiplier of human wealth. We Americans have borrowed against the world's store of inexpensive energy in the same way that we borrowed against the illusory equity in our homes, and we have used that energy leverage to construct a way of life that will cease to be sustainable the moment we can no longer cover our monthly payments.

ON A SHELF IN MY OFFICE IS A PILE OF RECENT BOOKS about the environment which I plan to reread obsessively if I'm found to have a terminal illness, because they're so disturbing that they may make me less upset about being snatched from life in my prime. Near the top of the pile is *Out of Gas: The End of the Age of Oil*, by David Goodstein, a professor at the California Institute of Technology, which was published in 2004. "The world will soon start to run out of conventionally produced, cheap oil," Goodstein begins. In succeeding pages, he lucidly explains that humans have consumed about a trillion barrels of oil (that's 42 trillion gallons), or what Goodstein estimates to be about half of the earth's total recoverable supply; that a devastating global petroleum crisis will begin not when we have burned the last drop but when we have reached the halfway point—an event known as peak oil, or Hubbert's Peak, after the geophysicist who first predicted it[7]—because at that moment, for the first time in history, the line representing supply will fall permanently through the line representing demand; that we will probably pass that point within the next few years, if we haven't passed it already; that various well-established laws of economics are about to assert themselves, with disastrous repercussions for almost everything; and that "civilization as we know it will come to an end sometime in this century unless we can find a way to live without fossil fuels."[8]

Extreme predictions have a history of turning out badly. In 1999, in a cover story called "Drowning in Oil," the British magazine *The Economist* warned that a barrel of oil, which then cost about ten dollars, might soon fall to five dollars, with potentially horrendous global consequences, including the decimation of oil companies' exploration budgets. "Cheap oil will also mean that most oil-producing countries, many of them run by benighted governments that are already flirting with financial collapse, are likely to see their economies deteriorate further," the magazine prophesied.[9] All this seems comical in retrospect—by the end of that same year, the price of oil had nearly tripled—but, at the time, the prospect of dramatically cheaper oil did not seem ridiculous, at least to the oil-market experts at *The Economist*. It's conceivable that worrying today about oil's increasing scarcity in the next few years will turn out to have been as foolish as worrying a decade ago about its overabundance. After all, between mid-2008 and the end of 2008 the price of oil fell by almost 75 percent, and, at many times in the past, people have been wrong about oil, or have been off in their timing. As Robert Bryce writes in *Gusher of Lies*, which was published in early 2008, the impending end of oil has been predicted many times: in 1914 by the U.S. Bureau of Mines (ten years); in 1939 by the U.S. Department of the Interior (thirteen years); in 1946 by the U.S. State Department (twenty years); in 1951 by the Department of the Interior again (thirteen years); in 1972 by the Club of Rome, in *The Limits to Growth* (twenty years); in 1974 by Paul Ehrlich, in *The Population Bomb* (twenty-five years); in the 1980s by the petroleum geologist Colin Campbell (before the end of that decade).[10]

In all these cases, the prognosticators failed to anticipate new oil discoveries and advances in the technology of finding and recovering petroleum, and perhaps similar developments will make Goodstein seem like a worrywart, too. It's certainly true that in the first years of this century high prices gave drillers an incentive to seek oil in places where they wouldn't have looked if the price had still been ten dollars a barrel, and that expensive new techniques enabled them to extract oil that once seemed hopelessly inaccessible and therefore wasn't included, even a few years earlier, in estimates of how much recoverable petroleum remained. Even so, the world's total supply, whatever it may actually be, is currently shrinking at the rate of about 3.5 billion gallons a day, while global demand can only rise. The world's population is expected to reach 9 billion by 2042, an *increase* equal to more than seven times the current population of the United States, or equal to the combined current populations of China and India. You don't need to have read the second chapter of your college economics textbook to guess what lies ahead, whatever the exact timing turns out to be.

One important fact about the economics of oil, emphasized by Goodstein and others but often overlooked, at least in media reports, is that the world will never actually run out, in the sense of pumping the final barrel out of the ground—in contrast to the moment, poignantly described by Jared Diamond in his book *Collapse,* when some anonymous Easter Islander cut down his island's last palm tree, bringing his culture to a permanent end.[11] At some point long before the earth has literally gone dry, extracting the remaining crude will cease to be economically rational, and production will essentially stop—either because, as Amory Lovins and others have predicted, increased energy efficiency and competition from solar, wind, and other renewables will have driven the price of oil so low that people will no longer bother with it (just as we no longer bother to hunt whales for fuel to light our houses), or because, as seems vastly more likely, finding and pumping petroleum will have become so expensive that the world will have switched almost entirely to burning more plentiful carbon-based alternatives and their derivatives— natural gas, coal, trash, wood, the petroleum extractable from so-called oil sands—and to making more use of nuclear power. Indeed, this second transition is already well under way all over the world, with environmental, economic, and geopolitical consequences yet to be fully reckoned. Supplies of those fuels are finite, too, and they are subject to the same peak effects as oil, but their current relative abundance will allow the party to go on for some time. A full transition to renewables will have to take place at some point, but we have a very long way to go. As of the end of 2007, the installed generating capacity of all the photovoltaics in the United States, both off-grid and grid-connected, added up to less than half of the generating capacity of the Hoover Dam.

Some economists dismiss the notion of catastrophic energy peaks. Steven D. Levitt and Stephen J. Dubner, in the revised and expanded (2006) edition of their best-selling book *Freakonomics*, write:

> What most of these doomsday scenarios have gotten wrong is the fundamental idea of economics: people respond to incentives. If

the price of a good goes up, people demand less of it, the companies that make it figure out how to make more of it, and everyone tries to figure out how to produce substitutes for it. Add to that the march of technological innovation (like the green revolution, birth control, etc.). The end result: markets figure out how to deal with problems of supply and demand. Which is exactly the situation with oil right now. I don't know much about world oil reserves. I'm not even necessarily arguing with their facts about how much the output from existing oil fields is going to decline, or that world demand for oil is increasing. But these changes in supply are slow and gradual—a few percent each year. Markets have a way of dealing with situations like this: prices rise a little bit. That is not a catastrophe; it is a message that some things that used to be worth doing at low oil prices are no longer worth doing. Some people will switch from SUVs to hybrids, for instance. Maybe we'll be willing to build some nuclear power plants, or it will become worthwhile to put solar panels on houses.[12]

This is an unusually conventional argument for Levitt, who calls himself, in his book's subtitle, a "rogue economist," and it doesn't address the central oil-peak contention, which is that, although demand for oil may rise at the rate of just a few percent each year, the economic consequences of declining reserves accelerate and reverberate as soon as new production can no longer keep up. As the world's experience in the past few years has shown, markets haven't yet come close to producing a replacement for oil—any more than they have found a way to eliminate disease or malnutrition, or to modify the weather so as to prevent floods and hurricanes, even though the economic incentives for doing those things have been high for millennia. The global recession that took shape in 2008 put time back on the oil clock, and OPEC actually reduced production in an effort to support prices, but the long-term outlook is unchanged. Oil comes out of a hole in the ground, and we set it on fire. It's not a clever but outdated invention of ours, like pie safes or black-and-white TVs, that we will replace with something cleverer as soon as the market determines that doing so is worth our while. The near certainty is that, for many years to come, what the market will replace oil with is not something better (such as nuclear fusion, which, at the very least, is decades or generations away) but something considerably worse (such as low-grade coal, China's main fuel, which makes oil's carbon footprint and pollution profile look demure), and that ordinary market forces, rather than leading us inexorably toward a golden future, will most likely entice us to compound our growing troubles by prompting us to invest heavily in the energy equivalents of patent medicines (such as shale oil and ethanol). Sometimes, the invisible hand goes for the throat.

It now seems clear that the main reason oil became so expensive for Americans in 2008 was speculation, even market manipulation. Another reason was that the dollars we were buying oil with had declined in value in comparison with other currencies. But those dollars were weak in part because we had spent so many of them on oil and other imports that the world outside our borders had more dollars than it knew what to do with, a condition that hasn't gone away. Other forces, unrelated to ordinary supply and demand—such as the form of market anxiety that the oil expert Daniel Yergin described, in testimony before Congress's Joint Economic Committee in 2008, as "shortage psychology"—undoubtedly contributed too, although shortage psychology, as it relates to oil, is not delusional; it's more like the fears of the paranoiac in the joke, the one who has real enemies. At any rate, the long-term trend is not in doubt. There is no brilliant Federal Reserve chairman, no stroke of monetary genius, no breakthrough strategy of the Group of Eight which can increase the volume of oil contained within the earth, or reverse the steadily increasing cost of bringing it to the surface and turning it into fuel. Market analysts, business journalists, economists, and other professionals whose business is thinking about oil often convey an eerie sense of unreality when they speak of it, since they refer to the act of removing it from the ground as "production," as though petroleum were an ordinary manufactured good, like shoes, or a simple commodity, like wheat, whose supply and price are driven up and down mainly by changes in human desire. Market turbulence obviously affects the availability of oil from one day to the next, but oil, unlike shoes or wheat, has an end point. Mere yearning cannot permanently increase its supply or solve the tremendous scientific, economic, and political challenges of making it irrelevant.

STARTING SOME HUNDREDS OF MILLIONS OF YEARS AGO, the remains of dead plants and other organisms began to accumulate in profusion on the bottoms of swamps and seas, became mixed with and were buried under various heavy sediments, and, as a result of geological and other processes occurring over millions of years, were subjected to tremendous pressure and heat, which eventually transformed them into the conveniently ignitable substances we call the fossil fuels: oil, natural gas, coal, and certain other, closely related, substances. In 2003, Jeffrey S. Dukes, who at the time was a postdoctoral fellow in biology at the University of Utah, calculated that every gallon of gasoline we burn today represents the transformed remnants of almost a hundred tons of prehistoric plant material—roughly the same quantity of biomass to be found in a forty-acre wheat field, including the stems, leaves, and roots.[13] On reporting assignments in 2007 and 2008, I visited the island of South Uist, in the Outer Hebrides, off the west coast of Scotland, and saw fields from which local residents had excavated peat, which they burned to heat their houses. Peat is a coal precursor, an intermediate phase between the mucky floor of a Carboniferous swamp and a gleaming seam of anthracite, and if you look closely at a drying slab, which resembles scorched sod, you get a much clearer sense of the genealogical link between green plants and power plants than you get from contemplating a can of motor oil. Fossil fuels truly are fossils. The energy they contain is solar energy that was captured and stored by living things and then transmuted, by time and geology, into exceptionally potent, compact, and easily transportable forms. Dukes estimated that the fossil fuels burned on earth in a single year, 1997, arose from a quantity of ancient organic matter equivalent to more than four hundred times one year's growth of all the plant and plantlike material on earth, including all the algae and plankton in the oceans.

An alternative theory of the origin of fossil fuels holds that they aren't fossils at all, but instead arose, and still arise, "abiogenically," from nonliving sources of hydrocarbons deep inside the earth. This idea, first proposed by a Russian chemist in the 1800s, has few adherents among serious scientists today. It is a favorite, though, of crackpots and credulous bloggers, and also of a good friend of mine, who is a global-warming skeptic and an unflagging environmental optimist, and who regularly e-mails me links to websites which claim either that glaciers are refreezing, or that ocean temperatures are rising not because of accumulating greenhouse gases in the atmosphere but because of heat escaping from undersea volcanoes, or that the world's supply of oil is, for all practical purposes, unlimited. I first read about the abiogenic theory in an article in *The Atlantic Monthly* in 1986, and was deeply gladdened by it—what a relief!—but the passing of two decades has not enhanced its plausibility. The simplest refutation is that the process it describes, if it isn't solely a wishful fantasy, doesn't occur on a scale sufficient to have made it detectable. There are no reports of exhausted oil fields quietly replenishing themselves.[14]

Coal has been burned as a fuel for tens of thousands of years, beginning in regions where chunks of it could be picked up off the ground. Serious underground mining began in Britain in the 1700s, and it genuinely changed the world: the Industrial Revolution could accurately be called the Coal Revolution, since the machines and factories and locomotives and steamships that proliferated in the nineteenth century were powered, both directly and indirectly, by coal. (The steam engine was invented primarily to pump water out of British coal mines.) Coal is an amplifier of human labor. In Britain, it transformed the toil of Welsh miners into the lights of London, the mechanized looms of Manchester, the steel mills of Sheffield, the freighters unloading cotton on the docks of Liverpool. Coal was crucial because by the late 1600s the British had virtually burned their way through what had once been a huge native endowment of timber, the earth's first plentiful ignitable fuel. The founding of Jamestown, in 1607, was intended partly as an energy-finding expedition: the scarcity of firewood at home had constrained the British glassmaking industry, and Jamestown's investors hoped that the colonists, by taking advantage of North America's seemingly limitless supply of combustible vegetation, would be able to manufacture glass there and ship it back to Britain.[15] And the Coal Revolution continues today; in 2005, coal accounted for more than a quarter of the world's energy consumption, a proportion that will rise during the next two decades and beyond.[16]

Oil has been known for about as long as coal, and was used for a number of purposes by various ancient peoples, in Mesopotamia, Egypt, and elsewhere. They collected it from natural seeps—places where underlying accumulations breached the surface, like the black ooze that made the Clampetts millionaires on *The Beverly Hillbillies.* (The La Brea tar pits, in Los Angeles, are a natural upwelling of asphalt, another fossil hydrocarbon, also known as pitch.) The first modern oil well is often said to have been a sixty-nine-foot-deep shaft drilled by a Colonel Edwin Drake on his farm, in Titusville, Pennsylvania, in 1859, but there are other contenders, in the United States, Canada, Europe, and Asia. Oil carries far more energy than coal—roughly 42.5 million BTUs per metric ton, versus 22.9 million, though the figures vary by type—and it inaugurated an even greater industrial revolution, the one that created the modern world and continues to shape and support it today. (Oil currently accounts for about 37 percent of global energy consumption.)

In the United States, oil wells were the Internet start-ups of the late 1800s and early 1900s. Relatives of my mother's lived in western New York state, near the town of Olean. They were mainly farmers, timber harvesters, leather tanners, and bankers, but the area was rich in oil deposits, and several of them made fortunes from petroleum, both directly and indirectly. ("Olean" means "oily," approximately, and the town was named, in the early 1800s, after a nearby oil seep, which local Indians had exploited, probably for centuries.) For quite a while, Olean was the world's largest producer, the Saudi Arabia of its day. The world's first major oil pipeline—six inches in diameter, 315 miles long—was laid by Standard Oil between Olean and

Bayonne, New Jersey, in 1881, and three more six-inch pipes followed quickly. They were buried a foot and a half below the surface, and during cold weather hot salt brine was pumped along with the oil to keep it fluid. The four pipes, at their peak, carried a total of 50,000 barrels a day, an immense volume at the time.[17] (Today, by comparison, the United States goes through 50,000 barrels of oil every nine minutes.) No Olean oil money seeped down to my parents or me, alas, but I was vaguely aware of it when I was growing up. My father was a stockbroker, and he wallpapered our downstairs bathroom with certificates issued by companies that had gone defunct, among them various oil concerns in New York and Pennsylvania in which my relatives had invested. Oil's recent record prices revived interest in some of those oil fields, which had been considered tapped out since around the time of the Great Depression, although the stock certificates, if they still exist, remain worthless.

The first truly transformative commercial product made from oil was kerosene, which began to be refined in the late nineteenth century and rapidly became America's preferred fuel for lamps. Kerosene distilled from petroleum replaced coal oil, which was manufactured by crushing and heating certain types of coal or coal precursors, and the two fuels, between them, dealt a catastrophic blow to the New England whaling industry, which had passed its own peak and therefore was doomed anyway. (This was bad for Nantucket but good for whales. In 1860, a California publication reported, "Had it not been for the discovery of Coal Oil, the race of whales would soon have become extinct. It is

estimated that ten years would have used up the whole family."[18])
Gasoline—which has a lower boiling point than kerosene, and
evaporates first when crude oil is fractionally distilled—was
viewed initially as a useless by-product, since it was too volatile
and explosive for use in lamps, and was discarded.[19]

In the early decades of the twentieth century, producing oil
was seldom much more complicated than sticking a tube in
the ground and collecting what spewed out. In Olean in 1920,
a significant oil discovery was made when lightning struck a
seemingly unpromising well and ignited an underlying natural-
gas deposit, proving that the well wasn't dry, after all.[20] Olean's
oil deposits actually lay fairly far below the surface, in comparison with subsequent discoveries in other parts of the country,
although they were still just a few hundred feet down. The cost
of oil was little more than the cost of moving it from one place
to another, and that remained true, all over the world, for decades. In the late 1990s, I interviewed a billionaire who had
built part of his fortune on natural gas leases in the Anadarko
Basin. "Oil is the cheapest liquid in the world," he said, dismiss-
ing any suggestion that we might be in danger of running short.
Crude oil, at that time, was selling for less than twenty-five cents
a gallon, so making an argument in opposition was futile. Even
today, petroleum products are still amazingly inexpensive, compared with other manufactured fluids. The next time you fill
your tank, go into the convenience store at the gas station and
try to find liquids on the shelves which are priced lower than the
gasoline you just pumped into your car. Not the coffee. Not

the soft drinks. Not the mouthwash or the shampoo. Not the bottled water. The marvel is not that oil costs as much as it does but that for so long it has cost so little. For most of oil's history as a fuel—for almost the entire twentieth century—it was essentially free.

Yet oil replaces an extraordinary quantity of human physical effort. A healthy, well-fed laborer, working eight hours a day, can produce an average output of something like 75 or 100 watts of useful work; a diesel-powered machine, operating at 40 percent thermal efficiency, can produce 75 or 100 watt hours of work from a few tablespoonfuls of fuel. If you could capture all the energy in a typical barrel of crude oil—42 gallons, 5.8 million BTUs—and convert it into forty-hour human workweeks, it would be the equivalent of several years' worth of one man's moderately strenuous manual exertion. I own a modern reprint of a book called *Homemade Contrivances and How to Make Them: 1001 Labor-Saving Devices for Farm, Garden, Dairy, and Workshop*, which was first published in 1897, before American agriculture had been transformed by fossil fuels. It contains illustrated instructions for building manure sheds, hog stickers, cattle stunners, fishing scows, mink traps, wagon seats, floodgates, clod crushers, hay presses, chaff forks, nests for egg-eating hens, and hundreds of other handmade and hand-operated implements, and it explains how to make fertilizer from animal bones, swamp muck, the scrapings of roadside ditches, the mossy surfaces and hard tussocks of swamp meadows, and other unlikely materials. To read the book is to be staggered both by the ingenuity of human beings and by the extraordinary amount of labor that used to go into saving labor, before the proliferation of machines powered by coal, oil, and natural gas. Nowadays, a dairy farmer who required a milking stool—if such a dairy farmer still exists—would be as likely to order one online ($27.95 at www.lehmans.com) as to spend an evening fashioning one from the bottom of a peach basket, a couple of old leather belts, a short length of wood, and various other items lying around the barn.[21]

Most of us think of our careers as the products of choice: I chose to be a journalist, you chose to be a lawyer, your friends and siblings and children chose to be plumbers and farmers and doctors and actors and carpenters and factory workers and truck drivers and college professors. In reality, though, our careers are mainly the products of fossil fuels. I am able to sit at a desk all day, staring at a computer screen and being distracted by e-mail, televised golf, and online bridge, because coal, oil, and natural gas are out there somewhere, doing my share of the heavy lifting. A manual laborer producing 100 watts' worth of physical work for eight hours a day, 365 days a year, would have an average work output of 33 watts; a typical North American, by comparison, consumes energy at an average rate of 12,000 watts. Most of that difference—between 33 and 12,000—is made up by fossil fuels.[22] In 1801, Thomas Jefferson, in his first Inaugural Address, said that the American wilderness would provide growing room for democracy-sustaining agrarian patriots "to the thousandth and thousandth generation." He didn't foresee the interstate highway system, and his arithmetic was off, in any case, but the main reason that our world differs so dramatically from his is the discovery and massive industrial exploitation of coal, oil, and natural gas. The modern history of our species is the history of our ascent up what the naturalist Loren Eiseley called "the heat ladder."[23] If the earth had contained no convenient source of densely packed, readily storable, easily transportable, and quickly releasable energy, America today would necessarily be closer to the country that Jefferson envisioned, and more of us would be toiling in fields and forests and workshops and factories, and building our own hog-scalding vats and corn-fodder ventilators and jigs for putting points on wooden-fence pickets (assuming we hadn't already starved to death or succumbed to any of the innumerable health emergencies that petroleum-derived chemicals have permitted us to overcome), rather than worrying about the recent performance of our 401(k)s. Coal bested firewood as an inexpensive multiplier of economic productivity, and oil and natural gas bested coal. The fossil fuels have enabled us to massively leverage the strength of our bodies, allowing a single farmer to produce the harvest of many, and to produce it on less land, and to ship it farther away, freeing a steadily growing percentage of us to do something other than growing or finding food, and to think of our lives in terms of something other than simple survival. The fossil fuels are not merely conveniences, or raw materials, or sources of pollution, or dwindling natural resources, or tradable commodities, or foundations of great fortunes, or geopolitical chess pieces; they are the basis of modern life. Oil is liquid civilization. We are what we burn, and that's as true for the head of the Sierra Club as it is for you and me.

Another inescapable fact about petroleum is that the oil we burn today is different from the oil that my relatives burned in upstate New York a hundred years ago. A crucial consideration regarding any fuel is a calculation usually known as energy returned on energy invested, or EROEI—the ratio of the usable energy provided by a fuel to the energy that had to be expended to produce the fuel in the first place. In this sense, oil in the early days was almost pure profit: a barrel of crude could be removed from the ground with the expenditure of just 1 percent or less of the energy that the barrel itself provided, a return on energy investment of 100:1. That ratio has fallen as the world's shallowest, most accessible deposits have been exhausted; the EROEI of Middle Eastern oil today is more like 10:1, while the EROEI of oil removed from deep-water deposits in places like the Gulf of Mexico is lower still. David Goodstein writes, "As we progress down the fossil fuel list from light crude oil (the stuff we mostly use now) to heavy oil, oil sands, tar sands, and finally shale oil, the cost in energy progressively increases, as do other costs."[24] All such calculations are subject to angry, irresolvable debate. (The EROEI of Middle Eastern oil looks worse if you include the cost of American military expenditures, in Iraq and elsewhere, that were intended at least partly to preserve U.S. access to those supplies,* or if you try to include some reckoning of the environmental costs of producing and using various fuels, and it looks better if you factor in the increasing energy efficiency of modern machines.) In addition, fossil fuels are by no means interchangeable; some aren't even fuels. "Shale oil" is a misnomer, since shale oil—of which the United States has vast deposits—isn't oil at all but is, rather, a dirty, difficult-to-extract oil precursor, which can be turned into a burnable fuel only with massive environmental disruption and a huge input of energy, giving it an EROEI that is probably much less than 1. What is indisputable is that the most familiar fossil fuels—oil, coal, natural gas—even at prices much higher than their current ones, remain economically alluring, in comparison with any likely alternatives. The most widely touted oil replacement in the United States, corn-based ethanol, has an EROEI of something between just over 1 and well under 1, depending on what you include and whose calculations you accept; hydrogen—which

---

*The war in Iraq can be thought of as the only wholehearted and fully funded element of ex-president George W. Bush's plan to meet America's future energy needs. In 2002, before the war began, Vice President Dick Cheney drew a direct connection between oil and the decision to invade Iraq: "Armed with an arsenal of these weapons of terror, and seated atop ten per cent of the world's oil reserves, Saddam Hussein could then be expected to seek domination of the entire Middle East, take control of a great portion of the world's energy supplies, directly threaten America's friends throughout the region, and subject the United States or any other nation to nuclear blackmail."[25] One possibly salutary result of the Iraq War is that it has disabused Americans of the fantasy, widely held before the invasion, that U.S. forces can simply sweep into an uncooperative country and take control of its petroleum resources. We were bound to try at some point, probably. Now we have.

doesn't occur on earth naturally in nontrivial quantities, and must be produced either by "reforming" natural gas or gasoline (an energy-intensive process whose direct by-product is the greenhouse gas carbon dioxide), or by using huge inputs of energy to split water molecules—is worse.[26] At a meeting of representatives of Habitat for Humanity a few years ago, I criticized Leadership in Energy Efficient Design (LEED), the popular architectural certification program of the U.S. Green Building Council (about which I'll have more to say in chapter 5) for giving an environmental credit to buildings that provide "alternative fuel refueling stations" for occupants' vehicles. After the meeting, a builder asked me why I objected to that credit, and I said, "Because there are no alternative fuels." That is still true. Honda began manufacturing a few hydrogen-powered cars in 2008, and news reports made it sound as though the only barrier to their broad release (other than their terrifically high cost) was the scarcity of hydrogen refueling stations. That's certainly a barrier, but a far bigger one is the unsolved problem of economically producing, delivering, and storing hydrogen in the first place.[27] Joseph J. Romm, an MIT-trained physicist, who served in the U.S. Department of Energy for five years in the 1990s and is the author of *The Hype About Hydrogen*, has said that hydrogen-powered automobiles will become feasible "not in our lifetime, and very possibly never"—a view common among physicists, if not among politicians or car-company marketing departments.[28]

One consequence of oil's falling EROEI is that oil has been

exerting peaklike influences on the world's economies for quite some time, as the cost of producing it has risen in comparison with the value of the energy it provides. It also means that as the world necessarily moves beyond easily extractable oil—and, therefore, becomes more dependent on more expensive oil and on fuels that are less efficient, in terms of the energy they yield— the fuel leverage that Americans have exploited for a century will inevitably decline, too. The fuels we use tomorrow will replace less labor than the fuels we have used up until now, both because they pack fewer BTUs and because more labor and energy go into extracting and processing them; and our dependence on them will nudge us not forward toward *The Jetsons* but backward toward *Homemade Contrivances*. This may not be entirely a bad thing; like most wealthy people, we have squandered much of our fortune, spending it in ways that didn't make us happier or healthier and creating unnecessary suffering for ourselves and others. Undoubtedly, there are innumerable ways in which we Americans can cut back dramatically without reducing the quality of our lives, and likely even increasing it. Indeed, there are many affluent or formerly affluent Americans who, in the past couple of years, have felt a sense of personal relief in their suddenly reduced ability to indulge in truly reckless consumer spending. But cutting back on fossil fuels isn't like cutting back on restaurant meals or trips to Paris; it's more like cutting back on oxygen or water. Dramatically upending the economic basis of entire societies doesn't usually turn out well for those societies. Unsettling ramifications extend in every direction. The *New*

York Times columnist Thomas L. Friedman has written about what he calls "The First Law of Petro-Politics," which states: "As the price of oil goes up, the pace of freedom goes down. As the price of oil goes down, the pace of freedom goes up."[29] There's that to brood about, too.

In 1987, the World Commission on Environment and Development, which had been established four years earlier by the United Nations, published an influential book, called *Our Common Future*, which summarized numerous international hearings on issues related to sustainability. "Growth has no set limits in terms of population or resource use beyond which lies ecological disaster," the book's authors wrote. "Different limits hold for the use of energy, materials, water, and land. Many of these will manifest themselves in the form of rising costs and diminishing returns, rather than in the form of any sudden loss of a resource base." This will certainly be the case with oil: we are currently in a race not with the actual end of petroleum but with the end of our ability and willingness to pay for what remains. "The accumulation of knowledge and the development of technology can enhance the carrying capacity of the resource base," the commission continued. "But ultimate limits there are, and sustainability requires that long before these are reached, the world must ensure equitable access to the constrained resource and reorient technological efforts to relieve the pressure."[30]

Two decades have now passed since the commission made this rather mild-sounding but actually quite dire warning, and world events have given us no reason to disbelieve it or to think that we have successfully followed its imperative. The summer of 2008 provided a sobering preview of the repercussions the global oil-production peak will eventually entail. The increasing scarcity of affordable oil amounts to an economic and environmental quadruple whammy. It directly and indirectly raises the price of nearly everything, including food, and it absorbs resources that might otherwise have supported more economically productive and socially desirable activities, and it encourages the widespread industrial conversion to more polluting, more climate-damaging, less energy-efficient fuels—"We are now getting into the dirtiest sources of oil anywhere," a Canadian energy consultant observed in 2006[31]—and, more broadly but less obviously, it unwinds the tremendous economic, societal, cultural, and other gains that the century-long abundance of cheap oil has made possible in the United States and other prosperous countries. Or is that a quintuple whammy? At any rate, the consequences are far-reaching, and the cost of refueling Ford Expeditions barely makes the list.

OIL'S DIRECT ROLE IN AMERICAN LIVES GOES FAR BEyond fuel. Plastic is made from oil—a fact that 72 percent of Americans in 2007 didn't know, according to an online survey[32]—and our world, increasingly, is made of plastic. Stephen Fenichell has written that when a French-American salvage expedition, beginning in the mid-1990s, retrieved items from the RMS *Titanic*, which sank in 1912, none of the thousands of recovered articles—"gold and silver wrist- and pocket watches, buttons, bracelets, jeweled necklaces, rings, tiepins and hairpins, gold pince-nez spectacles, leather goods, several hundred English coins, ivory combs, mirror cases, and hairbrushes"—included anything made of plastic, the first truly modern form of which, Bakelite, had only just been invented when the ship set sail. "That, if nothing else, shows how much times have changed," the secretary of the French Merchant Marine told a press conference during the salvage operation.[33] Since the sinking of the *Titanic,* plastic products have become so widespread that life without them can seem close to inconceivable. In 2008, when the supermarket chain Whole Foods stopped providing plastic grocery bags to customers, Natalie Angier, of *The New York Times,* wrote, "Bravo. Now tell me this: What am I supposed to line my garbage cans with? I always use plastic supermarket bags, and the Whole Foods ones were by far my favorites—roomy and springy enough to hold a lot of sodden waste without fear of breakage, always a plus when one is disposing of, say, fish skins or cat litter. So if I have to buy plastic bags by the box, that's better for the environment how? Forget about paper bags for this purpose. When we were growing up in the Bronx, my older brother recently reminded me, we lined our garbage cans with newspapers, a solution satisfactory to none but the roaches."[34]

Bags are the least of it. At the desk where I'm working at this moment, virtually everything within my reach is made at least partly of plastic and, therefore, at least partly of oil: computer CPU, keyboard, monitor, scanner, printer, printer stand, telephone, cell phone, digital voice recorder, compact disks, cables, cords, Velcro ties, labels, clock, calculator, headphones, chair, carpet, foam carpet pad, sneakers, wastebasket, wastebasket liner, camera, speakers, tape, pens, binders, file boxes, scissor handles, staple remover, coffee mug, desktop protector, padded envelopes, window air conditioner, latex wall paint, the lenses of my eyeglasses. There's a roll of dental floss in a pile of stuff near my printer; it has a sharp little metal tooth on the top, for snipping lengths of floss, but every other component—plastic container, plastic spool, nylon cord, wax coating, discarded package—was made from oil. The few items I can see in my office that don't contain plastic—books, magazines, papers, postage stamps, business cards, my wool sweater, the stain and finish on my wooden desk—contain inks, dyes, coatings, adhesives, and other elements that were manufactured from oil, and all of them were brought to me, many of them from the other side of the world, on ships and trains and trucks powered by oil. The toys in the playroom, the cosmetics in the bathroom, the drugs in the medicine cabinet, the hand lotion by the bed, the soap in the (plastic) kitchen dispenser, the packages in the cupboard, the cleaning supplies under the sink, the rain gear in the closet, the clothes on the clothesline, the clothesline itself: all were made from oil.

Oil is a cornucopian resource. A typical barrel of light crude, when refined, yields about nineteen gallons of motor gasoline, thirteen gallons of fuel oil (including diesel and home heating oil), four gallons of jet fuel, a gallon and a third of asphalt, and a miscellany of other astonishingly useful substances. Among the minor products of oil distillation are ethane and propane, which are further processed to create the synthetic polymers from which modern plastics are made. Those polymers arise first in the form of a powder-like substance, which is combined with other materials (catalysts, plasticizers, dyes), then melted, cooled, and cut into pellets the size of peppercorns. The pellets are shipped to fabricating plants, where they are melted again, then extruded, molded, injected, blown, and otherwise transformed into the products we see all around us.[35] When I was in grade school, my science class toured a local plastics factory—run by the father of one of my classmates—in which mountains of such pellets were turned into finished goods, mostly containers for household products. I went home that afternoon with a freshly injection-molded shampoo bottle and a pocketful of baby-blue pellets just like the ones from which I had watched my bottle being made.

Part of our ongoing environmental crisis involves what we do with our plastic possessions once we've finished with them—which is usually to throw them away, often within minutes of getting them home. (A third of all the plastic consumed in the United States is packaging.) In the 1990s, Curtis Ebbesmeyer, an oceanographer in Seattle, drew attention to an area in the Pacific Ocean, about a thousand miles off the coast of California, where circular ocean currents have formed a 10-million-square-mile vortex containing a relatively high density of floating bottle tops, beverage cups, volleyballs, G.I. Joe accessories, and other plastic detritus, now known as the Great Pacific Garbage Patch. Charles Moore has written:

The potential scope of the problem is staggering. Every year some 5.5 quadrillion (5.5 x 1,015) plastic pellets—about 250 billion pounds of them—are produced worldwide for use in the manufacture of plastic products. When those pellets or products degrade, break into fragments, and disperse, the pieces may also become concentrators and transporters of toxic chemicals in the marine environment. Thus an astronomical number of vectors for some of the most toxic pollutants known are being released into an ecosystem dominated by the most efficient natural vacuum cleaners nature ever invented: the jellies and salps living in the ocean. After those organisms ingest the toxins, they are eaten in turn by fish, and so the poisons pass into the food web that leads, in some cases, to human beings. Farmers can grow pesticide-free organic produce, but can nature still produce a pollutant-free organic fish? After what I have seen firsthand in the Pacific, I have my doubts.[36]

In 2008, a dead Cuvier's beaked whale, which had washed up on a beach on the Isle of Mull, was found to have twenty-three plastic bags or parts of bags in its stomach. They had gotten there because floating plastic bags look like jellyfish to creatures that eat jellyfish. The same year, a researcher at the Marine Biological Laboratory at Woods Hole, Massachusetts, hypothesized that the ingestion of certain petrochemicals released by the disintegration of plastic refuse in the oceans was causing a massive die-off among lobsters.[37]

Just as harrowing as the thought of what we do with petroleum-based plastics is the thought of what we will possibly do without them, once the inexpensive oil is gone. Will we go back to ivory combs and silver watches? Forsaking pantyhose, Under Armour, running shoes, and all our other oil-based garments for clothes made only of natural fibers—a strategy often recommended as environmentally sensitive—poses the same unhappy dilemma that promoting ethanol does: it diverts agricultural capacity (all those new cotton fields and rubber plantations!) away from the production of food. The world is suffering food shortages already; should vegans reconsider fur coats? More disturbing, plastic has played a major role in raising the life expectancies of the lucky inhabitants of the developed world. Sterile containers, disposable syringes, medical tubing, artificial corneas, the sanitary packaging that both protects and hugely extends the shelf life of many foods, thereby increasing the safety and efficiency of food production all over the world—how will we replace those? Americans' annual expenditure on plastic garbage bags exceeds the total annual expenditures, on everything, of nearly half the world's countries.[38] This is appalling, of course, but no one who, like Natalie Angier's brother, is old enough to remember life before Hefty bags would be eager to return to the era of reeking, leaking, maggot-infested garbage barrels.

Two of the earliest synthetic plastics—celluloid and rayon, both of which preceded Bakelite—are made not from oil but from cellulose, an abundant, naturally occurring compound found in green plants. (Cellulose gives plant stems and tree trunks their rigidity, and is the principal component of cotton.) Cellulose-based plastics, such as true cellophane, are largely biodegradable, at least in theory, but they also have numerous drawbacks, among them a tendency to catch fire easily, a characteristic they share with paper and firewood, which are also cellulose-based. Ping-Pong balls are still made of celluloid (and are spectacularly combustible), but in most other once popular applications (movie film, sports shirts, billiard balls, clear food wrap) cellulose has largely, and for good reason, been supplanted by other materials, the vast majority of which are derived from petroleum. In addition, some environmentalists and others are already eyeing cellulose as a potential replacement for corn as a source of ethanol. Cellulosic ethanol, as it's called, is difficult to make and may never actually be commercially viable—Robert Bryce has called it "the vaporware of the energy sector"[39]— but if fuel scientists ever do solve the considerable challenges involved in producing it economically manufacturers will have to compete for arable land with the producers of cellulose-based plastics. They will also have to compete with the producers of food.

Plastics can be made from plant materials other than cellulose. In 1942, Henry Ford received a patent for a method of automobile construction which employed plastic body panels made from soybean oil. "Plastic parts have many advantages in that they produce a quiet body, may be molded to exact sizes, . . . may be readily replaced in case of accident, and result in a lighter construction," Ford wrote in his patent application, which he submitted in 1940.[40] Soybean cars were a dead end, but Ford's interest helped to turn soybeans into a major American cash crop, something they had never been before. Ford was a friend and champion of George Washington Carver, who, in addition to finding hundreds of commercial uses for peanuts, devised many uses for soybeans, including the creation of fuel. Ford endowed the George Washington Carver Laboratory, in Dearborn, Michigan, a research institution whose activities included supplying the city's Ford Hospital with soy milk, a product that Ford himself had helped to develop. (Ford's career can be viewed as a campaign to eliminate large domesticated animals. Having made the horse obsolete, with the automobile, he had now set out to eradicate the cow.)[41]

Pea starch and high-fructose corn syrup, among other unlikely substances, may also turn out to be useful as natural sources of polymers that can be turned into commercially viable plastics. But these possibilities create the same discouraging dilemma that all plant-based petroleum replacements do: we can't eat our crops and wear them, too, especially if we are also depending on them to fuel our cars. When you read about current research involving the potential environmental significance of various agricultural products—exciting discoveries involving biofuels, bioplastics, vegetation-based carbon sinks, and increased food supplies for the world's rapidly growing human population—you sometimes get the feeling that no one is talking or listening to anyone else. Multiple competing interests have big plans for the same limited resources. This exasperatingly complex problem affects more than petroleum. The August 2008 issue of *Scientific American* contains an editorial that advocates addressing Africa's rapidly expanding food crisis by inaugurating "a program that not only delivers better seeds to African farmers but also devotes still more assistance to support improvements in soil, irrigation, roads, and farmer education. Then, when necessary, we should use remaining aid money to buy either hybrid or genetically modified crops grown in African soil for local distribution. The U.S. farm lobby will howl in protest, but this action will be the best way to work toward putting African bread on African tables."[42] The editorial makes a compelling case for that plan. Yet an article elsewhere in the same issue indirectly makes a compelling case against it. Peter Rogers, in "Facing the Freshwater Crisis," argues that agriculture in arid countries places unsustainable pressure on already limited water supplies, and that the solution is for the developed world to ship food to those countries. "A kilogram of wheat," he writes, ". . . takes about 1,000 liters of water to grow, so each kilogram 'contains' that quantity. Export-trade shipments of wheat to an arid country means that the inhabitants will not have to expend water on wheat, which eases pressure on local [water] supplies."[43] Undoubtedly, both analyses are correct. But the proposed solutions are mutually exclusive.

AMERICANS CONSUME APPROXIMATELY 850 MILLION gallons of crude oil per day. That's a bit less than a quarter of the world's production, and it works out to about 2.8 gallons per person. Actually, when you consider everything that Americans currently do with oil—commuting to work, heating homes, operating farm machinery, manufacturing plastics and pharmaceuticals, transporting goods, flying all over the globe on a whim, fueling the world's largest and most mobile military force* — 2.8 gallons of crude oil a day doesn't seem as though it could possibly be enough. Is that really *all* we use? And the truth is that we actually do use quite a bit more, indirectly, since a portion of the oil consumed by the rest of the world is consumed in the course of making things to sell to us. One reason that Chinese oil consumption has risen so dramatically in recent years is that we Americans have been importing vast quantities of Chinese manufactured goods, the production and exportation of which consume large quantities of oil and other fuels, especially coal. (The writer Patricia Marx, before visiting Shanghai on a reporting assignment in 2008, asked a friend who lives there whether she wanted Marx to bring her anything from the United States. "No need," the friend replied. "Chances are, any item you would bring from there is made here."[45] Equally significant, our purchases from the Chinese have gone a long way toward creating the wealth that has allowed large numbers of Chinese to acquire

---

*The U.S. Department of Defense, which used approximately 117 million barrels in 2006, is the largest single oil consumer in the world. It uses more oil per capita than all but three of the world's countries. Every one-dollar increase in the price of a barrel of oil costs the Pentagon an extra $130 million a year.[44]

the oil-based luxuries that we ourselves have enjoyed for decades, including automobiles.) No people on earth have had a bigger impact on global oil consumption than we have, and no developed country of significant size—other than Canada—has created an economy, a culture, a national infrastructure, and a general way of life that are more disastrously oil-dependent than ours. Approximately two-thirds of the oil consumed in the United States and Canada is used for transportation, mainly automobiles, a use for which there is currently no remotely attractive fuel substitute, while in Europe oil's main uses are for heating and power-generation, which are less dependent on oil and, at least in theory, less dependent on other fossil fuels.[46]

Fuel use by automobiles is a murky topic even for people who, you'd think, ought to know better. In 2004, an executive of a national environmental organization told me that her understanding was that if all Americans would switch to hybrid automobiles the United States would no longer need to import oil—a claim that also turns up fairly often in blogs and cocktail-party conversation. Here's the arithmetic on that one: the United States has proven recoverable oil reserves of something less than 30 billion barrels. If Americans decided to remove themselves from the world oil market and use only domestic oil, at our current rate of consumption, until the final drop was gone, those 30 billion barrels would last less than four years. (In reality, there would be no way to get all that oil out of the ground in four years, but just assume.) Doubling the average fuel efficiency of all American motor vehicles—a much larger gain than could actually be expected from even a complete, instantaneous conversion to hybrids—would add only about a year to that figure, and drilling in the Arctic National Wildlife Refuge, which may contain an additional 10 billion barrels, might add another. The American Petroleum Institute, which is the principal trade association of the American oil-and-natural-gas industry, estimates that total U.S. territorial reserves—including "undiscovered technically recoverable" oil in places that are currently inaccessible—are much higher, 112 billion barrels. This is a number that not even the oil companies take seriously, but even if you assume that it isn't a gross exaggeration 112 billion barrels is still less than fifteen years' worth at the current U.S. rate of consumption.* From a purely economic point of view, the United States, rather than inveighing against the importation of foreign oil, should have spent the 1990s buying as much foreign oil as possible and pumping it into the ground within our borders, so that we could have sold it in 2008 for ten or fifteen times what we paid for it. That would have been some investment.[47]

Even so, the popular idea that pumping more oil from American territory would help the United States become "energy-independent" from the rest of the world is just a fantasy. The oil market is global. That means that, as a friend of mine put it

---

*The API says 112 billion barrels is "enough oil to power 60 million cars for 60 years"—which is another way of saying that it's enough to power the actual number of passenger vehicles in the United States (about 250 million) for something like fifteen years. And this, remember, is a number from the outer limits of plausibility.

recently, the world's oil is "sold into, and bought out of, a single big barrel." Closing our borders to imports (thereby turning ourselves into petroleum locavores) would hurt only us, by ending our access to the big barrel. Ripping apart the National Arctic Wildlife Refuge or any other ecologically precarious area in order to extract the oil underneath it would slightly increase the amount of oil for sale in the world, but at a tremendous potential environmental cost for those areas. As the economist Robert Reich has said (in explaining why there is no significant advantage to American oil consumers—as opposed to American oil companies—in lifting various bans on off-shore drilling in U.S. coastal waters), "We take the environmental risk, but we'd have to share the negligible price gains with Chinese consumers and every other user around the world." It would be better for us to think of the National Arctic Wildlife Refuge as our desperate, last-ditch national oil reserve, and save it for truly desperate times—like, maybe, the day when we finally run out of plastic grocery bags.

Most environmentalists will tell you that the steep increases in U.S. gasoline prices leading up to mid-2008 were a good thing for the environment because they created a long-needed economic disincentive to wastefulness, as well as an inducement to develop alternatives. One reason that we Americans burn as much gasoline as we do—44 percent of the world's total consumption—is that fuel has always cost us very little, in relation to historical prices and to the prices paid by most of the rest of the world. Until 2007, when the cost of oil rose precipitously and stayed there for the first time in many years, gas prices in the United States, after being adjusted for inflation, were actually lower than they had been in the late 1970s and early 1980s, and weren't that much higher than they'd ever been, even during the period, in the middle decades of the twentieth century, that many people think of as having been the golden age of driving. During the summer of 2008, the price of regular unleaded at my local gas station edged over five dollars a gallon, the most I've ever paid in the United States. Yet many environmentalists will argue convincingly that U.S. gas prices, even at those levels, were still disastrously low, and that a truly enlightened national energy policy would include fuel-tax increases or carbon-tax charges that would, at the least, bring American prices into line with those in Europe, where a gallon of gasoline is much more heavily taxed and typically costs more than double what it does here.* But notice that increasing the fuel efficiency of a car is mathematically indistinguishable from

---

*Raising fuel taxes in the United States is probably politically impossible—almost all the public clamoring and congressional posturing in 2008 were for doing the opposite—and with a crippled economy it might even be calamitous. But high energy taxes are a good idea, generally, for reasons that go beyond their direct environmental impact. For example, increasing the tax on motor fuel, by forcing down U.S. petroleum consumption, would constitute a diversion of wealth from oil producers, including OPEC, to local, state, and federal treasuries in the United States. Such a tax increase could be made "revenue neutral" by pairing it with a compensating reduction in other taxes—perhaps including payroll taxes, which are highly regressive.

lowering the price of its fuel; it's just fiddling with the other side of the same equation. If doubling the cost of gas gives drivers an environmentally valuable incentive to drive less, then doubling the efficiency of their cars makes that incentive disappear. Getting more miles to the gallon is of no benefit to the environment if it is accompanied by an offsetting increase in driving—and the standard reaction of American drivers to decreases in the cost of driving, historically, has been to drive more.

How people react to increases in energy efficiency is a question of environmental significance. In 1865, the British economist William Stanley Jevons, in a book called *The Coal Question*, observed that coal consumption had increased, rather than declined, following the introduction of steam engines that used less of it—a phenomenon still known as the Jevons Paradox.[48] Whether you find it paradoxical may depend on how closely you observe your own behavior. Internet usage is analogous. The evolution from dial-up connections to cable connections, which are hundreds of times faster and therefore enable users to complete in seconds tasks that used to take minutes or hours, did not lead to a steep drop-off in connection time; on the contrary, that tremendous gain in efficiency helped to foster an even huger increase in usage. Downloading a webpage today takes just a fraction of the time that it did only a few years ago, and yet—paradoxically?—we stay connected longer and find more webpages we want to download. Energy use has followed the same pattern. In 2005, in a book called *The Bottomless Well*, Peter Huber and Mark P. Mills explained why: "Efficiency may curtail demand in the short term, for the specific task at hand. But its long-term impact is just the opposite. When steam-powered plants, jet turbines, car engines, light bulbs, electric motors, air conditioners, and computers were much less efficient than today, they also consumed much less energy. The more efficient they grew, the more of them we built, and the more we used them— and the more energy they consumed overall."[49] This has always been true in the past, and it's hard to think of reasons that it might not continue to be true in the future. The less costly a useful task becomes, the more we tend to do it; the less expensive an air conditioner is to operate, the longer we leave it on. This is true even in communities in which environmental awareness is unusually high. During the past decade, the 4,300 residents of the Danish island of Samsø have dramatically reduced their dependence on fossil fuels, and they now produce, from a variety of renewable sources, including wind turbines and straw-burning power plants, more energy than they use—a notable accomplishment. Yet their total energy consumption has not fallen. Elizabeth Kolbert, who wrote about Samsø in *The New Yorker* in 2008, quoted an exasperated-sounding resident, who said, "We made several programs for energy savings. But people are acting—what do you call it?—irresponsibly. They behave like monkeys." Kolbert cited, as an example, the fact that "families that insulated their homes better also tended to heat more rooms."[50]

The Jevons (and Samsø) Paradox doesn't mean that inefficient cars are a good thing. But it does suggest that increasing energy efficiency of our machines will not, in the long run, deliver the kind of environmental benefits that are often predicted for it. When the price of a barrel of oil first reached $110, in early 2008, National Public Radio ran a news story in which a reporter spoke with a young woman who owned a Chevrolet Tahoe, which she said she used mainly to drive (alone) between home, work, and school. "I kind of regret getting me an SUV," she said, and added that she hoped the price of gas wouldn't rise above $3.50 because, if it did, "I'm just going to have to go on the bus, I think."[51] She laughed when she said that, as though it were an amusing absurdity, but she had pinpointed the dilemma. Moving her from a car to a bus would be a good outcome for the environment, not only because it would shrink her personal fuel consumption and reduce, by one car, the outward pressure that causes inefficient suburbs to metastasize, but also because it would help to supply the critical mass of users on which successful, cost-effective transit systems depend. By contrast, moving her from her current gas guzzler to a more sensible car would be a relative environmental loss, because her reduced fuel cost would merely relieve the economic discomfort that had caused her to think about public transit in the first place. In 2008, Joseph B. White, *The Wall Street Journal*'s excellent automobile columnist, looked back more than thirty years, to the period following the oil embargo of 1973. "When oil prices soared in that era," he wrote, "interest in electric cars, windmills, solar-heating panels and other petroleum alternatives accelerated. When conservation and new oil discoveries caused oil prices to collapse, the economic justification for expensive, immature oil-replacement technology collapsed as well, and it was a skip and a jump to the age of the SUV."

In the long run, driving is still driving. Improving the fuel efficiency of cars may, in the short term, somewhat slow the rate at which the world exhausts its endowment of petroleum, but in the end making driving less expensive merely encourages people to drive more. Better cars alone, no matter how many miles they get to the gallon, can't shrink mankind's carbon footprint or move us closer to solving the ultimately unavoidable problem of what comes after cheap oil. I have met self-satisfied Prius owners who acted as though every mile they drove were a gift to humanity; for them, switching to a more fuel-efficient car merely enabled them to stop thinking of their own automobile use as an environmental issue. There's an episode of *The Simpsons* in which Ed Begley, Jr., who is one of the more sanctimonious members of Hollywood's growing legion of actor-environmentalists, makes a cameo appearance in which he good-naturedly mocks himself by saying that the teensy, nonpolluting car he's driving is "powered by my own sense of self-satisfaction."[52] Increases in fuel efficiency can actually be bad for the environment if they aren't accompanied by cost increases, tax hikes, or policy measures that lead consumers to feel they have no choice but to find or create alternatives to hundred-mile solo car commutes.

This is why the environmental example set by dense cities is so important. One certain consequence of oil's increasing scarcity and cost is that Americans in coming years are going to have to figure out how to be far less dependent on cars—which is another way of saying we're going to have to figure out how to live more like New Yorkers and European city dwellers. As you might expect, that isn't going to be easy, especially for people like me. But New York City's environmental record is long and well documented, and it is actually quite encouraging, since we already understand the technology that makes it possible. Manhattan's example is also instructive because per-capita automobile use there has always been extremely low, regardless of what was happening with the price of oil.

Three

There and Back

Oil's true significance in American life began to emerge in the early 1900s, with the rising popularity of the automobile. People nowadays tend to think of cars as an urban or suburban phenomenon, but the most enthusiastic early customers were residents of sparsely populated areas like mine: nearly two-thirds of the first million buyers of the Model T lived on farms or in small towns, and by 1920, when driving was still relatively uncommon in cities, more than half of all farm households in the Midwest had a car. Troop mobilization for the Second World War took many young American men not only abroad but across state lines for the first time, and when they returned home after the war they retained a taste for flexible transport. Car ownership spread rapidly throughout the United States in the late 1940s and early 1950s, and it has continued to grow. A critical threshold was crossed in the late fifties, when the parents of the Baby Boom generation—who were now living in suburbs with broods of young children, and were facing the same transportation dilemma that Ann and I faced when we moved out of New York City—began to buy second cars. (In 1949, only 3 percent of American families had owned more than one.[1]) That trend continued, with families eventually adding third and even fourth and fifth cars, and in 2001, for the first time, the number of automobiles in the United States exceeded the number of licensed drivers.[2] Meanwhile, U.S. consumption of gasoline rose from about 11.5 million gallons per day in 1920, to 43 million in 1930, to 110 million in 1950, to 243 million in 1970, to 304 million in 1990, to approximately 390 million gallons per day today.[3]

The car's role in our ongoing environmental crisis goes far beyond the direct burning of gasoline, however. Cars don't just use energy themselves; they also raise energy consumption in all forms and in all categories, with the usual environmental consequences, by enabling people to live in ways that are unavoidably inefficient. Cars have propelled our centrifugal expansion away from centers of density, and as we have spread across the countryside we have drawn behind us an increasingly expensive, complex, and energy-dependent train of civic infrastructure. The consequences touch everyone, and they do so in a multiplicity of ways that don't necessarily seem obvious until you begin to think about them. Richard M. Haughey, in a booklet titled *Higher-Density Development: Myth and Fact*, which was published by the Urban Land Institute in 2005, wrote:

> Not only do local governments absorb much of the cost of more and more local roadways, profoundly longer water and electrical lines, and much larger sewer systems to support sprawling development, they must also fund public services to the new residents who live farther and farther from the core community. These new residents need police and fire protection, schools, libraries, trash removal, and other services. Stretching all these basic services over ever-growing geographic areas places a great burden on local governments. For example, the Minneapolis/St. Paul region built 78 new schools in the suburbs between 1970 and 1990 while simultaneously closing 162 schools in good condition located within city limits. Albuquerque, New Mexico, faces a school budget crisis as a result of the need to build expensive new schools in outlying areas while enrollment in existing close-in schools declines.[4]

I grew up in Kansas City. When I was a boy, in the 1960s, my father and I and various friends of his and mine used to go camping on an undeveloped piece of property on the outskirts of the city. That area seemed like wilderness to me then—when we hiked into the woods to find a tall tree for a rope swing, I felt as though we had slipped over the edge of the civilized world— but during the past forty years the city has radiated so far beyond it that today our old camping site, which is now part of a suburban residential neighborhood, could almost be considered the center of town. Every new outlying subdivision, every new corporate campus, every new shopping mall has pulled the city's mantle of infrastructure farther from the core, and, in doing so, has hugely increased duplication and waste, along with per-capita energy consumption and production of pollutants, including greenhouse gases. And all of that low-density growth has been driven and sustained by cars.

The energy inefficiency of individual automobiles, in other words, is a far less important environmental issue than the energy inefficiency of the asphalt-latticed way of life that we have built to oblige them—the sprawling American landscape of subdivisions, parking lots, strip malls, and interstate bypasses. The critical energy drain in a typical American suburb is not the Hummer in the driveway; it's everything the Hummer makes possible—the oversized houses and irrigated yards, the network of new feeder roads and residential streets, the costly and inefficient outward expansion of the power grid, the duplicated stores and schools, the two-hour solo commutes. Suburbanites who trade down from an SUV to a hybrid may cut their personal automotive-fuel consumption by half or two-thirds, but they have done nothing to address the far more fundamental and intractable environmental problem, which is that automobiles have enabled us to create a way of life that cannot be sustained without automobiles. This is why, in the long run, a car's fuel gauge is far less significant, environmentally speaking, than its odometer. In the same way that life in Manhattan is inher-

ently energy-efficient, whether or not residents consciously try
to conserve, life in the suburbs and beyond is inherently wasteful, no matter what kinds of cars the residents park in their garages, or how assiduously they swap incandescent lightbulbs for
compact fluorescents.[5] It's miles traveled, not miles per gallon,
that make the critical difference. A sprawling suburb is a fuel-
burning, carbon-belching, waste-producing, water-guzzling,
pollution-spewing, toxin-leaking machine, and, unlike a Hummer, it can't easily be abandoned for something smaller and less
destructive. We've spent a century erecting our way of life. Now
we must reconfigure it.

AMERICA'S SUBURBAN DIASPORA WAS BY NO MEANS AN
unanticipated consequence of the rise of the automobile. Henry
Ford, the great anti-urbanist, saw the car specifically as a tool for
destroying what he viewed as the curse of population density. In
1922, he wrote, "We shall solve the City Problem by leaving the
City. Get the people into the country, get them into communities where a man knows his neighbor, where there is a common-
ality of interest, where life is not artificial, and you have solved
the City Problem. You have solved it by eliminating the City.
City life was always artificial and cannot be made anything
else. An artificial form of life breeds its own disorders, and
these cannot be 'solved.' There is nothing to do but abandon the
course that gives rise to them."[6] Ford's stated goal was to use cars
to create the thing we now call sprawl, yet there was nothing

sinister about his way of thinking—which was widely shared then and, for that matter, is widely shared now. Nor was there anything sinister about Ford's underlying vision of the relationships between people. The writer Michael Pollan, who is the author of two excellent, thought-provoking books about food, *The Omnivore's Dilemma* and *In Defense of Food*, and who couldn't be more different, philosophically, from Henry Ford, touched on a similar theme while discussing energy in a recent magazine article. He wrote, "Think for a moment of all the things you suddenly need to do for yourself when the power goes out—up to and including entertaining yourself. Think, too, about how a power failure causes your neighbors—your community—to suddenly loom much larger in your life."[7] Pollan's subject is really the same as Ford's, the social bonds between individuals. For Pollan, though, cheap energy is the problem, while for Ford it was the solution. The difficulty, for the future, is that they're both right.

The architect Frank Lloyd Wright was a prominent early evangelist of automobile-based population dispersion. Among the most remarkable designs of his remarkable career was a series of concepts he created for Gordon Strong, a wealthy Chicago businessman, who had bought large tracts of land on and around Sugarloaf Mountain, in northern Maryland, about thirty-five miles northwest of Washington, D.C. Strong and his wife, Louise, wanted to develop the mountain as a tourist destination for motorists from Washington and Baltimore, and in 1924 Strong asked Wright to prepare a design for what came to be called the

Gordon Strong Automobile Objective. The principal feature was a ziggurat-like conical building, to be erected on the mountain's summit—a sort of inverted (and enlarged) version of Wright's Guggenheim Museum, which was still two decades in the future. Wright meant for the building to "complete" the mountain by adding a car-friendly man-made pinnacle. The idea was that visitors would drive up Sugarloaf's wooded slope, then continue in their cars across a bridge and ascend the building itself by means of a spiraling external roadway, which would wind around the outside of the structure. Wright called the building "the natural snail-crown of the great couchant lion" of the mountain. Inside the spiral would be restaurants, hotel rooms, and other features, which in various conceptions included a parking garage, a dance hall, a theater, and a planetarium. The spiraling roadway was to be paved partly with glass block, so that sunlight (and the moving shadows of passing automobiles) would be projected into the building's interior. At the very top of the structure was to be a steel tower, probably for mooring dirigibles.[8] Strong rejected the concept as excessively artificial—a decision that infuriated Wright. "I have given you a noble 'archaic' sculptured summit for your mountain," he wrote to Strong in 1925. "I should have diddled it away with platforms and seats and spittoons for introspective or expectorating businessmen and the flappers that beset them."[9]

Wright, like Ford, saw the car as a weapon against traditional urban life: it was a defining element in his evolving vision of an American utopia, which he called Usonia. Wright developed these ideas over several decades and described them in detail in his book *The Living City*, which was published in 1958 and incorporated two earlier volumes.[10] "To look at the cross section of any plan of a big city is to look at something like the section of a fibrous tumor," he wrote. "In the light of the space needs of the twentieth century we see there not only similar inflamed exaggerations of tissue but more and more painfully forced circulation; comparable to high blood pressure in the human system. Think of the big towns you know; then try to imagine what modern mobility and new space-annihilating facilities, even now, are doing to them!"[11] Cars (and personal helicopters) would enable Usonians to place a more comfortable distance between themselves and their neighbors, freeing them from the typical big-city streetscape, which Wright called "a vast prison with glass fronts."[12]

As was also the case for Ford, Wright's alternative to the traditional city was sprawl. Wright's notion of "organic architecture," whose aim is to make buildings seem to have arisen harmoniously from their surroundings and whose basic tenets are still deeply embedded in the thinking of American architects, including the ones who view themselves as green, is actually an argument for maximizing negative environmental impacts. Much of *The Living City* is devoted to Wright's specifications for a model planned community, called Broadacre City, which he had first presented in 1932. It had a population of about 4,000 and covered four square miles, of which each resident family's allotment was one acre. There were few buildings taller than a story or two, and there was no local public transit, since virtually all human movement was to be by means of individual motorized vehicles, which Wright also designed. (Wright's cars had two huge rear wheels and a single, spherical front wheel.) The ultra-low-density spacing of the residential real estate would permit Usonians to express their individuality through architecture, obviously an important issue for Wright, while also limiting unwanted contact with others. Wright described his vision as "the more natural life of the small town fruitfully expanded in the country,"[13] and he foresaw a patchwork of Broadacre Cities extending from coast to coast. He even created detailed plans for reconfiguring highway interchanges, to keep traffic moving in all directions without interruption. Broadacre City, like most utopian schemes, was more about the present than the future. The car, Wright realized, had changed everything, by eliminating the need for proximity and, potentially, turning every American place into an Automobile Objective. Broadacre City was suburbia as imagined by an engagingly cuckoo (though brilliant) monomaniac.

The same vision of unrestrained mobility shaped the country's first zoning regulations, which were also made both possible and necessary by the rise of the automobile. Zoning is the practice of sequestering like civic uses in discrete zones, or districts: single-family residences here, apartment buildings there, stores over there, and factories off in the distance, with everything connected by roads. This concept was by no means entirely new, since people all over the world had made similar divisions, both formally and informally, for centuries. (In 1291, the government of Venice moved that city's entire glassmaking industry to the island of Murano, in the Venetian Lagoon, both to limit the danger that the glassmakers' furnaces would touch off a catastrophic citywide fire and to make it less likely that the secrets of Venetian glassmaking would be stolen by outsiders.) But the internal combustion engine—combined with the crucial fact that so much of the raw land in the United States remained untouched, and therefore could be developed to suit automobiles, something that was less true in Europe—had made genuine isolation feasible, by enabling people to separate daily activities by greater distances than could easily be covered on foot, with the help of horses, or by existing networks of trains and trolleys. The result, in the United States, was the end of what Peter Newman and Jeffrey Kenworthy, the authors of *Sustainability and Cities: Overcoming Automobile Dependence,* have called the Transit City, and the beginning of the Automobile City. "With the availability of cars," they wrote in 1999, "it was not necessary for developers to provide more than basic power and water services since people could make the transportation linkages themselves. As this 'ungluing' process set in, the phenomenon of automobile dependence became a feature of urban life. Use of an automobile became not so much a choice but a necessity in the Auto City. And as automobile dependence became dominant, the Automobile City began to lose much of its traditional community support processes."[14] Newman and Kenworthy are Australian, and their country's predicament is very similar to that of the United States, because in Australia, too, the most significant waves of human settlement arose relatively late and, therefore, arrived in cars.

New York City adopted, in 1916, the first U.S. zoning ordinance, but New York was an unusual case because Manhattan, in particular, had already been developed so intensively that the impact of many central zoning concepts was less dramatic than it would later be in other municipalities, where population densities were far lower and undeveloped land was more abundant. The zoning idea was very much in the air in the United States in those years. In 1919, Herbert Hoover, who was then Woodrow Wilson's secretary of commerce, and who was an evangelical advocate of zoning, wrote, "Someone has asked, 'Does your city keep its gas range in the parlor and its piano in the kitchen?' That is what many an American city permits households to do."[15]

Spreading out is the concept at the heart of virtually all traditional zoning ordinances. Edward M. Bassett, who has been called the father of American zoning and who was the principal author of New York City's original ordinance, worried that subways would undermine this beneficial effect of increased automobile use, by encouraging people to live too close together. A 1998 report on the hidden costs of gasoline says that traditional zoning plays "a significant role in the inefficiencies of low-density development by creating two distinct infrastructures in place of the traditional multipurpose town or city. With the home and the workplace separated, often by long car commutes, two well-serviced developments are created with duplicate retail, service, and parking institutions: the bedroom community and the office park."[16] Standard zoning regulations prohibit or sharply limit almost every characteristic that Jane Jacobs celebrated as the irreducible ingredients of urban vitality, and that the Sierra Club has identified as tools for reducing or reversing sprawl. Zoning tends to fully separate residential and commercial uses, to move buildings farther apart and farther from streets and sidewalks, to force low-density development by limiting building height and lot coverage, and to require the creation of oversized parking facilities, which move buildings still farther apart, usually making them inaccessible to anyone who isn't driving. The increased personal mobility provided by cars had already made density unnecessary, as Henry Ford and Frank Lloyd Wright had both realized; now zoning rules, often, made density legally impossible.

Zoning regulations are so commonly assumed by their supporters to be bulwarks against thoughtless property development that it is easy to overlook the fact that the rules often actually undermine the community qualities they are thought to protect. I'm the chairman of my town's zoning commission, and over the past fifteen years I've had many opportunities to ponder that contradiction. Among the most picturesque and locally cherished sections of my town are the old village green and the oldest of our three small business districts. Both arose long before cars or the concept of zoning, and both have numerous fundamental features that standard zoning regulations explicitly forbid in new construction: the lots are small, the buildings are close to each other and close to the road, public parking is scarce, and commercial and residential uses are mixed, seemingly arbitrarily. These features—which are among the defining characteristics of the village centers of all picturesque old New England towns (and which, on a vastly larger scale, are among the defining characteristics of Manhattan and all the old cities of Europe)—are the ones that make visitors stop their cars and take photographs. Yet nothing remotely resembling either neighborhood could be built in our town today under our basic regulations for business or residential districts, because, as is the case in most towns, our standard rules regarding lot coverage, building setbacks, parking spaces, and the separation of uses, among other things, prohibit them. Nevertheless, those local residents who are the most concerned about preserving the remaining vestiges of our town's historical character are often the most impassioned in speaking against any proposed regulation change that would relax those requirements in those same districts. They tend to view "restrictive" regulations as invariably protective, even when the features that are restricted are the features they are hoping to encourage and preserve.

This same disconnect between ends and means is manifestly evident in a neighboring town, much larger than mine, where reckless low-density growth has proceeded at a truly remarkable pace during the past few decades. That town has a charming village green, consisting of a spacious public lawn surrounded by old buildings, many of them dating to at least the nineteenth century. In recent years, the town has spent millions of dollars to enhance the appearance and economic vitality of that central area, where many old businesses have failed, in the hope of preserving it as a living focus of community life. At the same time, though, the town has encouraged precisely the kind of commercial sprawl that has marginalized the village green—and now the state, with broad local support, is spending tens of millions of dollars to double the width of the town's principal traffic artery, a project whose main purpose is to carry an ever-increasing volume of cars out to the massive parking lots of Wal-Mart and Home Depot and Super Stop & Shop and the like, which have proliferated along the road in areas that were still mainly farmland just twenty or thirty years ago, and which are directly responsible for the economic and social decline of the old central village.

It is our cars that stand between us and solutions to our gathering energy nightmare. And it's easy to see why. Cars have defined our culture and our lives. A car is speed and sex and power and emancipation, and receiving a driver's license remains one of a very few adult rites of passage that never disappoint. (A friend of Ann's and mine found her son, who was ten, moping in the kitchen one day, and asked him what was the matter. He said, "I wish I had a car.") Even after almost forty years as a driver, I still look forward to long car trips—and not because my car is anything special. Some of my favorite driving experiences took place in the succession of minivans we owned when our kids were little. Since the kids have grown up, my wife and I have both moved on to (slightly) more sensible cars, but I still feel the pull. Not long ago, I saw a multipage magazine advertisement for the redesigned Chrysler Town & Country minivan, in which the second row of seats can be swiveled backward, toward the third, and I suddenly felt a surge of longing: I could picture my golf buddies playing cards on the little table that pops up between the rearmost rows of seats as I happily sped us toward some distant golf course, and I found myself wondering whether I couldn't think of some way for a middle-aged empty-nester to justify buying a 6,000-pound six-passenger minibus that gets sixteen miles to the gallon in non-highway driving. When I was in college I felt sorry for classmates who had grown up in New York City and didn't know what it was to cruise aimlessly along suburban streets on a summer evening, radio blaring, elbow out the window, girlfriend in the seat beside you—and I still feel sorry for them. A car makes its driver a self-sufficient nation of one. It is everything a city is not.

Most of the car's most tantalizing charms have turned out to be illusory, though. Driving, by helping us to live at ever greater distances from one another, has undermined the very benefits that it was meant to bestow. Ignacio San Martín, an architecture professor and the head of the graduate urban-design program at the University of Arizona, told me, "If you go out to the streets of Phoenix and are able to see anybody walking—which you likely won't—they are going to tell you that they love living in Phoenix because they have a beautiful house and three cars. In reality, though, once the conversation goes a little bit further, they are going to say that they spend most of their time at home watching TV, because there is absolutely nothing to do." In 2008, I flew to Atlanta late on a clear spring night. The Atlanta metropolitan area is the biggest sprawl bomb in the United States. It covers approximately 8,500 square miles, making it more than half again as large as my entire state, and 94 percent of its 5.2 million residents travel to and from work by car.[17] I was struck, as I looked down at the lights on the ground during the plane's descent, not only by how astonishingly far the city's suburbs have spread beyond the old city center but also by the fact that the largest features visible on the ground were vast, empty parking lots, most of them illuminated so brightly, even at midnight, that you could have delivered babies in them. Streets and parking spaces are the air inside the sprawl balloon. We have allowed them to steadily push us outward, farther from one another, and farther, even, from all our Automobile Objectives. We have done so willingly, confident that we were getting the better end of the bargain. But what a terrible price we have paid, and have yet to pay, for our liberation from the city.

AN OBVIOUS WAY TO REDUCE RELIANCE ON CARS IS TO turn drivers into public-transit passengers. This is easy to advocate but challenging to accomplish, because the same forces that have made increased transit use urgently necessary in the United States have also made it extraordinarily difficult to pull off. In 1960, 22 percent of Americans traveled to work by public transit or on foot, while 64 percent commuted by car; by 2000, transit users and walkers had fallen to 8 percent of the total, while drivers had risen to 87 percent—and more than three-quarters of those drove alone.[18]

In May 2008, as the price of gasoline was rising toward four dollars a gallon, Clifford Krauss, of *The New York Times*, reported that transit use was growing everywhere. "Mass transit systems around the country are seeing standing-room-only crowds on bus lines where seats were once easy to come by," he wrote. "Parking lots at many bus and light rail stations are suddenly overflowing, with commuters in some towns risking a ticket or tow by parking on nearby grassy areas and in vacant lots." Krauss reported that transit use in Denver during the first quarter of the year was 8 percent higher than it had been during the same period the year before, and that gains of "10 to 15 percent or more" had been recorded in cities in the West and South, "where the driving culture is strongest and bus and rail lines are more limited."[19] This sounded, at first reading, like very good news, but the changes that Krauss described were far less impressive than his account made them seem. As Paul Krugman pointed out in the same newspaper a few days later, "fewer than 5 percent of Americans take public transit to work, so this surge of riders takes only a relative handful of drivers off the road."[20] In most parts of the country (and even in Europe), transit has actually been stagnant or in decline—and even as record fuel prices stimulated increases in bus and rail use, the real gains in U.S. ridership were extremely small. (And they disappeared a few months later, when gasoline fell below two dollars.) That's because high fuel prices and public will alone are not enough to bring about truly meaningful increases in public-transit use in cities that have been developed in such a way as to make efficient transit impossible.

One of the few areas of the country to have experienced real, sustained growth in transit use over a significant period of time is New York. (Ridership on buses and subways in New York City grew by 29 percent between 1996 and 2000, and by 8.8 percent between 2003 and 2007.[21]) One result of that growth is that the city's Metropolitan Transportation Authority and Department of Transportation today account for nearly a third of all the transit passenger miles traveled in the United States and for nearly four times as many passenger miles as the Washington Metropolitan Area Transit Authority and the Los Angeles County Metropolitan Transportation Authority combined. New York's subway system is the third busiest in the world, after those in Tokyo and Moscow, and its routes encompass almost half of all the subway stops in the United States.[22] New York City's buses carry 842 million passengers a year—more than the combined total of the next three largest American bus systems, those in Los Angeles, Chicago, and Philadelphia.[23] Why has New York succeeded where others have failed?

New York City looks so little like most other parts of America that urban planners and environmentalists tend to treat it as an exception rather than a model, and to act as though Manhattan occupied an idiosyncratic universe of its own. Many contemporary writers on public transit don't even mention New York, or mention it only in passing, on the assumption either that a big dirty city has no place in a thoughtful discussion of green transportation systems, or that the only forms of transit worth talking about, from an environmental point of view, are light-rail systems in the Pacific Northwest and ethanol-powered buses in Brazil. But New York's public transit system is, far and away, the most successful in the United States, and the elements that have made it successful are the elements that are required to make transit successful anywhere. When transit truly works, it works for the same reasons that it does in New York; when it fails, it fails because the region it serves is too little like New York to make it feasible. New York, therefore, has more to teach an urban planner, in terms of the fundamental factors that determine the true value of any sincere effort to reduce American reliance on automobiles, than do places like Portland or Boulder.

Those fundamental factors were identified in 1977 by Jeffrey Zupan and Boris Pushkarev, of New York's Regional Planning Association, in a book called *Public Transportation and Land Use Policy*.[24] Zupan told me, "We put numbers on it, which—at the risk of sounding boastful—was really groundbreaking at the time." Since then, Zupan, Pushkarev, and others have produced many more such calculations, demonstrating incontrovertibly that the most significant factor in determining the viability of any transit system—far more significant than fare levels or demographics or willpower or anything else—is population density.

"The basic point," Zupan told me, "is that you need density to support public transit. In all cities, not just in New York, once you get above a certain density two things happen. First, you get less travel by mechanical means, which is another way of saying you get more people walking or biking; and, second, you get a decrease in the trips by auto and an increase in the trips by transit. That threshold tends to be around seven dwellings per acre. Once you cross that line, a bus company can put buses out there, because they know they're going to have enough passengers to support a reasonable frequency of service."

Zupan continued, "People often say they want public transit but then aren't willing to zone their area to achieve the density needed to support it. Then they complain to the government or the bus operator, How come you're not running a bus in my neighborhood? Well, they've got half-acre lots—what do they expect? Or they build their office building five hundred feet back from the road and put a parking lot next to it, so that a worker in a car can park by the front door, while anyone getting off a bus has to negotiate a muddy lawn, maybe not even with sidewalks, just to get to the building. Under such conditions, who's going to use transit? You can't have successful transit if you create an environment that doesn't support it."

Transit has a multiplier effect on the reduction of automobile use. Zupan and Pushkarev, in a 1980 study, estimated that every mile traveled by rail users in the metropolitan areas of New York, Chicago, Philadelphia, San Francisco, Boston, and Cleveland replaced four miles driven in cars, based on a comparison with a group of large American cities that lacked rail systems—a measurement known as transit leverage. John Holtzclaw, in a 1991 study of San Francisco, concluded that the city's rail system has an even higher transit-leverage ratio, of nine, meaning that every mile traveled on a train eliminates nine miles traveled by car.[25] Peter Newman and Jeffrey Kenworthy, in *Sustainability and Cities*, identified several likely reasons for the leverage effect:

- If a good transit option becomes available, then people and businesses adjust by locating nearer to the line; thus transit shortens travel distances.
- People taking transit often combine several journeys in one—for example, picking up groceries on the way home from work, which in a car-based suburban setting would likely mean separate car trips.
- Households that switch to transit often give up one car and thus have less car use because the choice of using a car is less available.
- Transit users often find that the habit of walking or biking to stations flows into the rest of their lifestyle.[26]

Transit leverage might better be thought of as density leverage, because much of the benefit arises from the fact that placing people close enough together to support efficient transit also encourages walking, biking, and other forms of non-automotive

getting around—what Holtzclaw has described as "the syner-
gism of public transit and high density mixed-use communities."[27]
In Manhattan, potential destinations are so close to one another
that walking is often a traveler's first choice, no matter what
other forms of transportation are theoretically available.

A second key to successful public transit is a lack of palatable
alternatives. New Yorkers don't ride the subway because they're
more enlightened or more environmentally aware than other
Americans; New Yorkers ride the subway because owning and
driving a car in the city is almost ridiculously disagreeable.
Curbside parking is scarce, and parking lots are shockingly expensive and often inconveniently situated or hard to find. Most
important of all, the average speed of crosstown traffic in Manhattan is little more than that of a brisk walker. At certain times
of the day, in fact, the cars on the side streets in midtown move
so slowly that they appear almost to be parked. Most people,
including most New Yorkers, view such clogged streets as an
urgent environmental problem, since the cars seem to just sit
there, spewing exhaust. But traffic jams like those actually generate environmental benefits, because they urge drivers (and cab
riders) either into the subways or onto the sidewalks. New York
is surely the only city in the America where you are likely to hear
someone say, "We'd better take the subway—we don't have time
for a cab." Making that cab ride more attractive than the subway,
by reducing the congestion on the streets, would be a loss for the
environment, not a gain.

Proponents of new or expanded transit systems often act as though the only significant keys to their success were infrastructure and determination—that if cities would build decent transit systems and residents would make a commitment to use them everything would be fine. But this has been proven, repeatedly, not to be true. A city can build the most attractive light-rail line ever conceived, and then spend millions urging residents to use it, but if the local population is too spread out to be served efficiently and cost-effectively by transit, and if driving remains a plausible alternative, then transit never achieves the ridership levels that were predicted for it—and no amount of browbeating, public-service advertising, or federal spending can change that. You can't reduce automobile use by a meaningful amount simply by lecturing people that riding the bus is better for the earth, or better for their soul. And that's true anywhere. As committed as New Yorkers appear to be to walking and riding the subway, it wouldn't take much to turn them into car owners. If every Manhattan apartment came with a free garage parking space, Manhattanites would have cars, too.

Probably the most celebrated modern transit success story in the world is that of Curitiba, Brazil, a city that, beginning in the 1970s, under the direction of an extraordinarily forward-thinking mayor, Jaime Lerner, adopted an innovative public-transit system based on buses. The American environmentalist Bill McKibben, who has been one of many notable promoters of Curitiba as an urban-development model, has written:

The buses move faster not because they have bigger motors but because they were designed by smarter people. Sitting at a bus stop one day, Lerner noticed that the biggest time drag on his fleet was how long it took passengers to climb the stairs and pay the fare. He sketched a plan for a glass "tube station," a bus shelter raised off the ground and with an attendant to collect fares. When the bus pulls in, its doors open like a subway's, and people walk right on (or, in the case of wheelchairs, roll right on). . . . In 1993 Curitiba added another Lerner innovation—extra-long buses hinged in two spots to snake around corners and able to accommodate three hundred passengers. Five doors open and close at each stop, and on busy routes at rush hour one of these behemoths arrives every minute or so; twenty thousand passengers an hour can move in one direction. There is a word for this kind of service: subway.[28]

McKibben is surely correct that Curitiba's bus system includes many features that ought to be copied by transit authorities everywhere, and that its example proves that, in urban areas that can be configured to support it, an innovative bus system can provide much of the speed and efficiency of heavy rail, at a very small fraction of the infrastructure cost. But there is another chapter to the Curitiba story, told less often. In the early years of Curitiba's transit transformation, the city was in deep financial distress, which Lerner's master plan helped to reverse, and those economic gains were amplified by the dramatic resurrection of the entire Brazilian economy, which used to be a disaster

but in recent years has been so robust that the country's boom-
ing oil-and-gas industry has had trouble filling thousands of
high-paying jobs. The result is that Curitiba today is a far more
prosperous place than it was when it first began to attract the
admiration of environmentalists, and the city's prosperity has
been accompanied by the usual modern ills. Curitiba, despite its
enviable bus system, now has the second-highest per-capita rate
of automobile ownership in Brazil, and it is known as "the De-
troit of South America" because its local industries include five
major automobile manufacturing plants. The city's five principal
bus routes, which radiate like spokes from the city center, were
intended to concentrate residential density along their lengths.
This they have done, but in doing so they have served as propa-
gators of suburban growth and have thereby undermined the
benefits they were meant to foster. The environmentalist Richard Register, in the 2006 edition of his book *Ecocities*, writes, "As
[Curitiba's] five high-density arms reached farther out into the
countryside, diverging ever more from one another, the area of
land between the arms grew much faster than the area of the
arms themselves. At some point in the city's recent history,
instead of the open space between the arms remaining open
space, low-density and much more automobile-dependent development between those arms began growing faster than the
transit arms themselves."[29] Curitiba has a swell transit system,
but as the impediments to car ownership have fallen, and as the
city's horizontal growth has increased the perceived usefulness of
driving, car ownership has grown.

This vicious cycle lurks just below the surface of all increases in prosperity and population. Los Angeles is at or near the top of almost everyone's list of the examples of automobile-dependent development, but L.A. is actually quite dense, as American cities go, with an average concentration, inside the city limits, of just over 8,200 people per square mile, or nearly 13 people per acre. This is fairly close to Zupan and Pushkarev's critical transit threshold of seven dwellings per acre—and it exceeds the density of many developments that have been promoted as examples of New Urbanism, or Smart Growth—yet only a microscopic percentage of Angelenos travel to work in anything but a car, and, largely because of the separation of uses mandated by local zoning regulations, there are very few parts of the city where transit, bicycling, or walking are feasible as regular means of getting around, no matter what the price of gasoline. Uninhibited car use invariably undermines the noblest of environmental intentions—always, everywhere.*

In the early 1900s, Los Angeles, like many other American cities, actually had a thriving streetcar system—a variety of the type of transit that, nowadays, is usually referred to as light rail. But then, as automobiles proliferated, streetcar riders and street-

---

*Las Vegas is another example of a city that is relatively dense—almost all its residential-building lots are very small—yet that can't support efficient public transit. As is true elsewhere in the United States, most commercial uses in Las Vegas are isolated in auto-dependent shopping centers and strip malls, accessible only by car. The one part of the city that functions like a dense urban core is the Strip, where hotel guests often walk from one casino to another.

car lines began to disappear. Beginning in the 1930s, National City Lines, a holding company created by General Motors, Firestone Tire, Philips Petroleum, and Standard Oil of California, acquired streetcar companies in L.A. and forty-four other cities and gradually shut them down, generally replacing them with buses manufactured by GM. "The L.A. Freeway era was born in the wake of this decision," Newman and Kenworthy have written. "It was not, however, a community decision—it was a commercial one, and illegal at that. National City Lines was found to have broken antitrust laws and fined $5,000. However, this commercial decision basically ended the Transit City era in the United States, particularly once the Federal Highway System began in 1956."[30] This story, which has been told many times and is a staple of environmentalist blood-pressure-raising sessions, shouldn't be contemplated by anyone with both a weak heart and negative feelings about large corporations. But the truth is that those old streetcar systems were doomed anyway. General Motors, along with the other American car manufacturers, has plenty of sins to answer for, but passenger counts on streetcar lines all over the country had been declining for years, and the National City consortium, even if it truly did set out to destroy the trolleys, had millions of willing co-conspirators in the growing legion of American automobile owners. There was no public force in the United States, in those days, that could have held back the death of the streetcar, even if anyone had had the will to try. As Sharon Feigon and David Hoyt wrote in 2003, "Development around the automobile has resulted in a type of urban form that now makes other mobility options inconvenient and uneconomical."[31] That was true then, too. Automobiles, virtually from the beginning, were destined to prevail.

This problem is obviously far more severe at much lower densities and in places where commercial and residential uses are sharply segregated. No transit system could ever function in my little town, even if money weren't an issue and all 4,000 residents agreed that riding the bus would be a good idea. Our houses and our likely destinations are scattered over such a broad area that creating practical, energy-efficient routes and schedules would be impossible. The part of town where I live is one of the oldest and, therefore, the most densely settled (relatively speaking), yet the street route from my house to the little commercial district in our town center, a distance of a bit more than a mile, directly passes fewer than two dozen houses—nowhere near enough to support even minimal, once-a-day service. What's more, about a third of the houses on the route are clustered at the very edge of the commercial district, where transit would be the least useful and the least likely to be used, since those distances are short enough for easy walking. And the farther you move from the center of town, toward my house and beyond, the more widely dispersed the dwellings and connecting roads become, making them increasingly difficult to serve with even minimal efficiency.

Actually, my town, like almost all American municipalities, does have a public transit system—its school bus service. Our public primary school serves 190 students, most of whom travel to school on the bus. The system employs five full-sized school buses, one minibus, and two vans, and each vehicle follows a circuit that covers about fifteen miles and takes forty-five or fifty minutes to complete, including the return trip to the bus terminal, which is about a mile from the school. This system represents close to the maximum in efficiency that could be expected from a small-town transit system, since the passengers all travel at the same time and many of them assemble in small groups at common collection points, reducing the number of stops. Even so, the system requires some passengers to spend forty minutes covering a distance they could have traveled in a few minutes in a car. If first-graders had driver's licenses, it wouldn't work.

My town is an extreme example, but similar problems haunt transit systems in municipalities with densities that, by comparison, seem vastly greater. Phoenix is the sixth-largest city in the United States and one of the fastest-growing among the top ten, yet its public transit system accounts for just 1 percent of the passenger miles that New York City's does. The reason is that Phoenix's growing population has been allowed to spread so far across the desert that no network of buses and trains could provide efficient service to all of it. (Between 1950 and 1990, the population of Phoenix grew by 819 percent, while its density fell by 63 percent.[32] Greater Phoenix today has a little more than twice the population of Manhattan, yet it covers more than two hundred times as much land.[33]) For transit to have a chance of working in a place like Phoenix, it has to be concentrated in areas that are dense enough to make it efficient, and it has to be thought of as a tool for further increasing the density of those areas, thereby reinforcing its own usefulness and reducing the city's overall dependence on automobiles. Transit, when used in that way, can help reduce the entropy of traditional real estate development, by turning its focus inward.

IN URBAN AREAS THAT ARE DENSE ENOUGH TO SUPPORT efficient public transit systems, officials often negate their own efforts to increase usage, by simultaneously spending huge sums to make it easier for people to get around in cars. When a city's streets or highways become crowded, for example, the standard response is to create additional capacity by building new roads or widening existing ones. Projects like these almost always end up making the original problem worse—while also usually taking years to complete and costing many millions of dollars—because they generate what transportation planners call "induced traffic": every mile of new, open roadway encourages existing users to make more car trips, lures drivers away from other routes, and tempts transit riders to return to their automobiles, with the eventual result that the new roads become at least as clogged as the old roads, though at higher traffic volumes, and the efficiency of transit declines. These negative outcomes are compounded by the fact that, in the short term, temporarily improved traffic flow reduces commute times for drivers on the expanded roadways, making it easier for people to justify building houses, shopping malls, and office buildings in formerly inaccessible outlying areas—and that, in turn, eventually makes all the original problems worse, as the places where commuters sleep and shop and work drift farther and farther apart, and new feeder roads are built to serve them.

The nation's second-largest public transit system, Chicago's, has shrunk in recent decades, as that metropolitan area's growing population and its expanding roadway network have spread ever farther beyond the city's core. This seeming paradox—that a growing city could undergo a retrenchment in transit—was explained by Jane Jacobs in 1961: "Increased city accessibility by cars is *always* accompanied by declines in the service of public transportation. The declines in transit passengers are always greater than increases in private automobile passengers."[34] Jacobs had been anticipated, six years earlier, by Theodore Kheel, a New York City labor lawyer and arbitrator, whose improbably long and distinguished list of career highlights includes his successful mediation of the major New York City newspaper strikes in 1962–1963 and 1978, and his designation by *The New York Times*, in 1988, as "perhaps the most influential industrial peacemaker in New York City in the last half-century."[35] In 1955, Kheel was asked by a group of nine private bus companies and two labor unions to study why transit ridership had declined precipitously in New York City. An article in the *Times* described Kheel's findings: "Having displaced the horse, the automobile is now inexorably replacing the mass transit facility," the article began. "In the process, a twin problem has developed: The more New Yorkers use private cars, the more transit traffic falls off and the more tangled street traffic congestion becomes. The result is frustration for both motorist and transit passenger."[36] Fourteen years later, Kheel, having been appointed by Mayor John V. Lindsay to head a committee charged with finding a way to slow growth in the city's transit fares, proposed imposing tolls on the free bridges and commuter parkways leading into Manhattan and doubling existing tolls on bridges that had them already, with a view to both generating needed transit funds and reducing the volume of car traffic entering the city. These suggestions angered Robert Moses, who by then was retired from his various official and unofficial positions of power but remained an influential consultant to the Metropolitan Transit Authority. Moses said, "You can't make rubber pay for rails"[37]— and his view prevailed, as it usually had. Moses' solution to almost any transportation problem had always been to build more roads—the standard first response, even today, of urban planners and state highway departments.

Building new public transit in the hope of reducing car traffic can be just as self-defeating as building new highways, if no other steps are taken to support it, because people seldom use transit unless they feel forced to, and unclogging roads, if successful, just makes driving seem more attractive, and the roads fill up again. New transit, if it is to succeed, has to be accompanied not only by population density sufficient to support it but also by a reduction in road capacity (to maintain a level of inconvenience that a significant number of drivers continue to find intolerable) or by a steady increase in the direct and indirect costs of using cars, or both—in effect, "induced transit," or what Jane Jacobs called "attrition of automobiles."[38] In a talk I gave to an environmental group in New York City in 2005, I suggested that the Nature Conservancy and similar organizations, which purchase parcels of unspoiled land in order to protect them from development, should consider buying downtown parking lots in big cities and erecting apartment buildings on them, since doing so would beneficially increase the population density of those urban centers while simultaneously making automobiles more annoying to use. I was looking for a laugh, but the point was serious. We Americans have shown consistently that we will make almost any sacrifice for our cars: we will pay horrifying prices for fuel and insurance, we will cut back on almost all other expenditures, including health care, we will endure extreme expense and inconvenience related to parking, and we will commute over distances that once would have struck almost anyone as inconceivable, if not insane. Building a gorgeous transit system is not enough to make people use it in large numbers; you also have to make the alternatives bleak, by increasing costs, impeding car traffic, and eliminating lanes and parking spaces. Transit use across the United States rose slightly in mid-2008, as the price of a gallon of gasoline rose above four dollars; it declined again a few months later, when gas fell by more than two-thirds.

Even in New York City, the complex relationship between traffic and transit is not well understood. A number of the most popular recent transportation-related projects and policy decisions affecting the city may merely have the effect, in the long run, of luring passengers into cars and away from public transportation by making driving more attractive. A good example is the recent reconstruction of Route 9A, Manhattan's West Side Highway—which runs along the Hudson River from the lower end of the Henry Hudson Parkway, at West Seventy-second Street, to Manhattan's southern tip, where it merges with the FDR Drive, its counterpart along the East River. The rebuilt highway, which was completed in 2001, is a traffic-engineering marvel: the roadway is grooved concrete, there are four lanes in each direction for much of its length, there are easy-to-use exits every couple of blocks below Fifty-seventh Street, and, on nice days, the views of the river can be spectacular. The new West Side Highway makes driving into the city (except during the very worst parts of rush hour) almost a pleasure, at least in comparison with what it used to be.

And that's the problem. The redesigned highway is so agreeable that it makes drivers less likely to take mass transit, form car pools, live closer to Manhattan, consolidate trips, or stay home. I consciously weigh this trade-off on the half-dozen occasions each year when I have to travel to the city, which is roughly ninety miles from where I live. Especially if I'm pressed for time, I will often decide to drive—something I went out of my way to avoid when the West Side Highway was being redone. Except during the morning and evening rush hours, the only part of the car trip that's slow is the very last bit, the five-block crosstown drive from the West Side Highway to midtown. That final segment, which is no more than a mile, can take a half-hour on a bad day. Yet the total car trip, largely because cars move more speedily along the rest of the route, now usually takes less time than the train. (The Metro-North train station nearest to my house is a forty-minute drive away, and the train takes almost two hours from there.) If the West Side Highway were narrower, more congested, and less convenient to use, as it was during the lengthy period when the roadway was being rebuilt, I would take the train every time. It seems absurd to suggest that people who design highways—people whose training and careers have been devoted to finding ways to move vehicles more expeditiously—should think about roads from an entirely different point of view, and apply themselves to making driving more infuriating, but from an environmental perspective that's what the city really needed, and what we all need. You can't make people drive less by giving them incentives to drive more.

Not long after the completion of the redesigned West Side Highway, part of the FDR Drive was rebuilt, too, and at the beginning of that project the federal government spent $139 million to erect a nine-block temporary bypass, called the Outboard Detour Roadway, so that users of the FDR (and residents of buildings on neighboring side streets) would not be discommoded while the work was under way. The bypass, which was built on concrete pilings sunk into the East River, was deemed necessary because that particular stretch of the FDR, between Fifty-fourth and Sixty-third Streets, had to be shut down completely during construction. (Part of the roadway runs under several apartment buildings and an esplanade.) In 2004, on the day before the bypass opened, Ian Urbina, of *The New York Times*, wrote that "if all goes well, the temporary roadway will preclude the hair-wrenching gridlock . . . that surely would have come had the city followed the conventional approach of simply closing down the lanes under construction, leaving drivers to fend for themselves."[39] This was the official rationale for constructing the bypass, and, as usual, the paramount consideration was the convenience of drivers. A hundred and fifty thousand automobiles use the FDR daily, and no serious thought was given to the possibility that some good might come from forcing their owners to seek alternative means of transportation. Meanwhile, the cost of the bypass alone was equal to 18 percent of the city's annual fare-box revenues from its entire bus fleet, or more than $55,000 a linear foot.

Public transit itself can be bad for the environment if it facilitates rather than discourages sprawl. In 2000, New York's Metropolitan Transit Authority added two new stations to the northern end of the eastern branch of the Metro-North Railroad, at the outer edges of its system, roughly eighty miles from Grand Central Terminal, and those extensions have made it easier for New York commuters to live even farther from the city's center, and have therefore encouraged the spread of new, non-dense residential development in previously isolated rural areas where all other travel is necessarily by automobile. Transit, in this case as in many similar ones, has actually been an engine of sprawl and, therefore, of all the environmental ills created and exacerbated by automobiles.

Atlanta's minimal rail transit system, which comprises just thirty-eight stops and forty-eight miles of track, has had a similarly negative effect. The Atlanta system has only two main lines—which run north-south and east-west, and intersect in the center of the city—and they have facilitated automobile-dependent suburbanization on the metropolitan compass points while doing almost nothing to reduce car use in the central city, where there are too few stations to make the trains a viable means of getting around. An acquaintance of mine, who works downtown but lives in one of Atlanta's newest, most distant subdivisions, told me that she had tried the train a few times because her car commute was hellish but had gone back to driving because, once she got to the city, she felt stranded without a car.

Transit, in order to be good for the environment and to reduce overall energy consumption, must be used to concentrate people in dense urban cores, rather than merely encouraging them to live farther from their jobs and other daily destinations. Building the long-discussed Second Avenue subway line, on Manhattan's East Side, is an environmentally sound project, because it will increase New Yorkers' ability to live without cars; building a bullet train between Penn Station and the Catskills (for example) would not be sound, because it would enable sprawling development to establish itself in a region that couldn't support it otherwise. Extending transit lines ever farther from urban cores—the standard practice in large metropolitan areas with existing systems—only makes the real problems worse. As Robert Cervero has written, "With the exception of the greater New York area (along the Metro-North corridor to Connecticut), relatively little land-use concentration or redevelopment can be found around US commuter rail stations—after all, the very premise of commuter rail is to serve the low-density lifestyle preferences of well-off suburban professionals who work downtown. Serving commuter trips almost exclusively also means that ridership is highly concentrated in peak hours, more so than any other form of mass transit service."[40] The purpose of transit should be to get people out of their cars, not to establish them in regions where the structural impediments to the worst kinds of wastefulness are few.

THE MOST TALKED-ABOUT TRAFFIC-RELATED ISSUE IN recent years, all over the country, has been congestion. Roads of all types and sizes, from twelve-lane urban expressways all the way down to one-lane country roads, are more crowded than they used to be, and traffic on the busiest thoroughfares is often reduced to a halting crawl. This is partly because there are so many more of us than there used to be (the U.S. population has more than doubled since 1950) and partly because we drive so much more than we used to.

To most people, traffic congestion looks like an ecological disaster. And it is one, but not for the reasons that people assume. Here's why: Traffic jams are not an environmental problem; they are a driving problem. If reducing congestion merely makes life easier for those who drive, then the improved traffic flow actually increases the environmental damage done by cars by raising

overall traffic volume, encouraging sprawl and long car commutes, and reducing the disincentives that make drivers think
twice about getting into their cars. Traffic jams are actually beneficial, environmentally, if they reduce the willingness of drivers
to drive and, in doing so, turn car pools, buses, trains, bicycles,
walking, and urban apartments into attractive options. Treating
congestion, rather than driving, as an environmental issue often
leads to transportation policies that, from an environmental
point of view, are flawed. Almost always, when traffic engineers
and others talk about reducing congestion what they are really
talking about is making traffic flow more efficiently, and that
means increasing the overall volume of cars—an obvious environmental negative.

Many cities have instituted so-called High Occupancy Vehicle
(HOV) programs on major commuter routes. These programs
offer special treatment to cars that contain multiple passengers,
usually by setting aside limited-access lanes, at least during rush
hours. HOV lanes have environmental value if they are treated
as a single element in a dynamic overall traffic-management system aimed at reducing total vehicle miles traveled, but they do
more harm than good if they are treated instead as a one-off tool
for improving the lives of people who would prefer to drive—as
is usually the case. The HOV system on I-84 near Hartford,
Connecticut, about an hour from my house, is a good example
of a bad idea. In each direction, it consists of a single lane, which
is isolated from the regular lanes by a broad and theoretically
uncrossable buffer of tarmac, and it's open to any car that con-

tains more than one occupant. Because there's just one HOV lane, passing is not a possibility, so before entering it drivers always scout for slowpokes up ahead, to make sure they wouldn't be better off in the regular flow. As is often the case with HOV lanes, any beneficial environmental impacts are undermined by the fact that the users include many vehicles that would have contained multiple occupants anyway—moms and kids on their way to school or soccer practice, elderly couples, families heading toward or returning from vacations—rather than former solo drivers who got together to save time or gas. Yet even when the system works the way it's supposed to, its beneficial impacts are limited because every genuinely doubled-up car that it attracts removes two cars from the regular lanes, making those lanes flow more readily and reducing, by two cars, the original incentive to switch. The only way to make an HOV system truly green is to adjust its configuration constantly, to maintain both a steady reward for carpooling and a continuing source of irritation for those who resist. If you move so many drivers into HOV lanes that solo traffic in the regular lanes begins to move faster, the original purpose is defeated, especially if HOV drivers get fed up and move back.* To make an HOV system effective, in terms of the environment, transportation departments have to be willing to gradually reduce the total number of lanes while either increas-

---

*Some HOV systems have allowed solo drivers of certain hybrid cars to use the lanes, even though the fuel advantage of hybrids disappears at highway speeds. A Prius is much greener in a traffic jam than it is on the open highway.

ing the proportion of lanes reserved for high-occupancy vehicles or increasing the number of occupants required to enter them. If anything, though, planners have tended to do the opposite. Early HOV systems often required three occupants; almost all of them now require just two, and many allow even that low barrier to be circumvented.

The HOV system in Atlanta—a city that has probably been the source of more bad transportation policy than any other in America—is characteristically absurd. In 2008, the Georgia Transportation Board, responding primarily to the unhappiness of automobile commuters, voted to pursue changing all forty-five miles of metropolitan Atlanta's HOV system into toll lanes that would be open to any drivers willing to pay the fee. The Transportation Board referred to these transformed lanes as High Occupancy/Toll lanes (HOT), even though high occupancy would not necessarily be rewarded. An article in *The Atlanta Journal-Constitution* explained: "Cars with only two occupants would almost certainly have to pay the toll, and maybe three-person cars too. Stretches of the HOV system are already so congested that if two-person cars stay in for free there wouldn't be any room for the toll-paying solo drivers."[41] In addition to repurposing existing HOV lanes, the Georgia plan calls for the construction of additional car lanes, perhaps to be financed privately, on which tolls would also be charged—and these lanes would be in addition to the other new lanes that Georgia is already building, using public funds. All these ideas, taken together, represent an extreme and almost comical example of the standard dog-chasing-its-tail approach of American traffic engineering: Georgia is struggling to keep up with the new traffic that it has induced in its misguided efforts to reduce congestion on its main commuter roads, and in doing so it is making its original problem worse and increasing the pressure on itself to adopt even more self-defeating measures later on. You can't solve a problem by aggravating the conditions that created it, even if doing so leads, in the short term, to the temporary alleviation of the most obvious symptoms.

Yet this is what traffic engineers are usually dedicated to doing. The goal is almost always to increase car volume, reduce commute time, increase average speed—in other words, to make car use more pleasant and dependable. These are gains for drivers but losses for the environment. In Los Angeles, in particular, maximizing car use has been elevated to a science. Tom Vanderbilt, in his fascinating 2008 book *Traffic: Why We Drive the Way We Do (and What It Says About Us)*, describes how L.A.'s Department of Transportation uses so-called real-time traffic signals, whose timing can be changed in response to local conditions, to increase car volume on the city's streets: "A study a few years ago by the DOT showed that the area containing real-time traffic signals reduced travel times by nearly 13 percent, increased travel speeds by 12 percent, reduced delay by 21 percent, and stops by 31 percent. Just by quickly alerting the DOT that signals have malfunctioned, the system squeezes out more efficiency. What the traffic engineers have done is add 'virtual' capacity to a city that cannot add any more lanes to its streets."[42] L.A.'s traffic engineers, in other words, have managed to induce additional car traffic in a city already seemingly operating beyond capacity. This is commendable only if the goals are to cram more cars into Los Angeles, increase the automobile dependence of an already automobile-dependent city, and make it possible for people to live still farther from where they work.

Making long drives seem shorter just keeps solo drivers wedded to their cars and exacerbates the underlying problem by enabling commuters to justify longer commutes. Moving cars out of regular lanes, furthermore, makes the regular lanes move faster, at least for a while, and therefore weakens the incentive to double up. Most American cities have the real problem upside down. Rather than allowing solo drivers to buy their way into high-occupancy lanes, these cities should be thinking of ways to make life harder for the least efficient automobile commuters—perhaps by requiring solo drivers to use congested, slower-moving lanes restricted to cars without multiple passengers, and charging them tolls to do so—with a view to eventually prompting those drivers to give up and join a car pool or take a bus. You can't reduce solo car use by rewarding drivers for continuing to drive alone.

Nor can you reduce car use unless reasonable alternatives, such as car-sharing programs and express commuter buses, are implemented on a sufficient scale and are actively supported. Yet even environmentalists tend to view commuting speed alone as the key measure of success or failure of almost any traffic-management scheme. Timothy Beatley, in a 2000 book called

Green Urbanism: Learning from European Cities, praises the HOT system in Orange County, California, by noting that "these lanes have been popular and have significantly reduced commute times and improved traffic flow"[43]—as though making automobile commuting less irksome were a worthy environmental objective.

Minneapolis is one of a number of American cities that have attempted to reduce the severity of rush-hour traffic jams by installing meters on freeway entrance ramps. These meters—which typically employ light signals or control gates to govern the rate at which vehicles are able to enter the flow of traffic—achieve their stated goal of reducing rush-hour highway congestion and, therefore, are commonly viewed as good for the environment. But they, too, actually undermine parallel efforts to persuade commuters to switch to less damaging forms of transportation and are therefore inherently wasteful. They do their damage by significantly decreasing peak-period travel times—by 10 percent in Atlanta and 22 percent in Houston, according to studies in those cities—and by invariably leading to increases in overall vehicle volume and, therefore, in the burning of gasoline. They also encourage sprawl by making longer car commutes seem reasonable. In Portland, ramp metering led to a 155 percent increase in average rush-hour travel speed, a boon to the residents and developers of distant subdivisions. In Minnesota, ramp metering increased overall traffic volume by 9 percent and peak volume by 14 percent—both desirable outcomes, according to the state's Department of Transportation, whose experience proves that it's possible to induce traffic without building new roads. The increase in traffic volume was accompanied by a corresponding increase in fuel consumption, of 5.5 million gallons during the six-week monitoring period. Despite this ominous result, the metering system was judged to be a success and an environmental boon—fuel consumption was referred to dismissively as "the only criteria category which was worsened by ramp metering," in a study summary prepared for the Minnesota Department of Transportation. The result was that the system was retained. The only significant program elements that were altered were ones that drivers had identified as bothersome to themselves.[44]

IN NEW YORK CITY, THE DEBATE ABOUT TRAFFIC CON-gestion has mostly concerned congestion pricing, also known as road pricing, which is the practice of charging drivers road-use fees in certain areas during the hours of the day when car traffic is the heaviest. In 2007, Mayor Bloomberg proposed a congestion-pricing plan for New York City by which drivers would have been required to pay eight dollars to enter Manhattan below Eighty-sixth Street between six in the morning and six in the evening. The mayor's office had estimated that this plan would raise $400 million in the first year—money that could be used to support public transit. But in 2008 the idea was killed by the state legislature, which had the final say. (Legislators were mainly worried that the plan would penalize drivers who couldn't afford the toll.)

Congestion pricing is basic economics. The idea is that if you have a sporadically scarce commodity, such as hotel rooms or space in automobile lanes, you can eliminate distribution bottlenecks by adjusting prices in counterpoint to variations in demand. Hotels do this by raising rates when travel is popular and lowering them when travel is not. That helps to smooth fluctuations in reservation rates, enabling the hotels to make better use of their existing rooms and to increase total revenues without building new capacity, much of which would end up being empty except during periods of peak travel.

The concept works the same with cars. Rather than attempt to eliminate congestion by building new roads or expanding existing ones, planners seek to make existing roads more efficient by imposing fees that are high enough to discourage significant numbers of drivers from traveling in the most popular places at the most popular times. This often does open up clogged streets—London's central business district is the main example that proponents usually cite—but the overall result is seldom a gain for the environment or for public transit. Increased efficiency is a good thing in the hotel business, economically speaking, but it isn't necessarily a good thing when it comes to the environmental impact of cars. If the result of congestion pricing is simply to spread traffic out, thereby maintaining or increasing total traffic volume while also making driving more pleasant for those who continue to do it, then its putative environmental benefits are fictitious. Generating funds for public transit is obviously a good thing—Robert Moses was worse than wrong when he said that rubber shouldn't be made to pay for rails—but attacking congestion by seeking to make traffic flow more smoothly has a serious downside unless all its implications are taken into account.

Before Mayor Bloomberg's New York City congestion-pricing plan was defeated, Transportation Alternatives, a nonprofit organization that sponsors a number of worthy initiatives related to walking, bicycling, and public transit, extolled it by saying, in part, "Mayor Bloomberg's proposal anticipates a 6.5% reduction in the number of vehicles entering Manhattan south of 86th Street. The effect will be even more dramatic during peak hours when an 11% traffic reduction will result in a 20–40% reduction in time lost to traffic delays. . . . Reduced congestion will benefit those who continue to drive, in the form of faster, more predictable commutes."[45] That was good news for drivers willing to pay the fee, but maybe not for anyone else. Time lost to traffic delays has an obvious cost—all those stalled commuters could be working at their desks or interacting with their children instead of fuming at other drivers, talking on their cell phones, or listening to audio books—but perceptions of productivity are among the factors that commuters weigh when they consider where to live and how to travel to work. Reducing congestion increases the productivity of solo driving, and that increases the incentive to drive—a bad result for the environment. In 1999, Newman and Kenworthy, citing a 1995 study published in the *International Journal of Urban and Regional Research*, concluded that "there is no guarantee that congestion pricing will simultaneously improve congestion and sustainability." They mentioned several ways in which congestion pricing can defy the expectations of its supporters, among them by causing motorists to "drive exactly as they always have if the congestion charge is covered by their firms (e.g., a majority of London's peak-hour commuters have company cars and perks)" and by causing them to "drive more as they shift to 'rat-running' through suburban streets to avoid congestion-priced streets."[46]

Advocates of congestion pricing usually argue that traffic jams waste gasoline, since cars stalled in traffic burn fuel when they're not moving.* That's true, but the energy waste and carbon output attributable to idling cars is vastly smaller than the energy waste and carbon output attributable to the overall transportation network, which generates waste both directly (by encouraging unnecessary driving) and indirectly (by encouraging

---

*The fuel waste of cars stalled in traffic jams is eliminated in gas-electric hybrids, which shut down their gasoline engines when they're not moving and are capable of idling and creeping along crowded freeways without burning fuel. Even nonhybrid cars could be given an "idle-off" capability—already used on gasoline-powered golf carts and some small cars outside the United States—which automatically shuts down engines at stoplights and in traffic jams and restarts them electrically, when needed, from a self-recharging battery.

forms of development that can be sustained only through huge new energy inputs and an ever-expanding web of energy-hungry infrastructure). There's nothing green about congestion pricing if, by distributing traffic more efficiently, it results in an overall increase in traffic volume, in extra miles driven by vehicles avoiding the fee areas, or in the increased attractiveness of driving for those individuals who don't mind paying.

That doesn't mean that charging drivers for driving is a bad idea. On the contrary, one of the most powerful forces propelling American sprawl has been the fact that the true cost of operating automobiles has always been hidden from those who truly pay it, and has been subsidized in numerous, complex ways that are probably impossible to decipher fully. A truly effective traffic program for Manhattan or any other dense city would impose high fees for all automobile access and public parking (thereby discouraging all car use and raising money to support transit) while also gradually eliminating automobile road capacity (thereby reducing total car traffic volume without eliminating the environmentally beneficial burden of driver frustration and inefficiency) and increasing the capacity, frequency, and efficiency of public transit. It is absurd, in New York, that the East River bridges still don't charge tolls, that the existing toll bridges are still relative bargains, that curbside parking in much of the city is free, and that the transit system is still burdened by the Moses-era conviction that free-flowing automobile traffic is a public entitlement, while subways and buses should have to fight for funding.

In 2007, following the defeat of Mayor Bloomberg's congestion-pricing plan, Theodore Kheel (the labor lawyer and mediator who in the 1950s identified the relationship between rising car use and declining transit described earlier in this chapter) proposed an alternative scheme, the main elements of which were charging cars sixteen dollars to enter Manhattan's central business district (double the Bloomberg fee), raising bridge and tunnel tolls, and reducing transit fares to zero. Free transit is actually an old proposal of Kheel's. He first suggested it in the mid-sixties, and the basic idea is simple: if cars are bad and buses and trains are good, why not make driving more expensive and use the proceeds to pick up transit's entire tab? Charles Komanoff, who was the research director and principal author of the formal version of Kheel's new proposal, which was released in January 2008, told me that he believes the Kheel plan has a chance with the state legislature because it balances the perceived penalty to drivers with an obvious boon to commuters who are willing to switch to transit. As the published proposal states, "Making transit free will remove, once and for all, the threat of fare hikes, and be an enormous boon for New Yorkers, particularly low-income residents, for many of whom free transit will bring a $20-a-week after-tax raise."[47]

It's not at all clear that free transit is a good idea, even if Kheel and Komanoff's budget can be made to balance. Among the many fascinating and apparently hardwired quirks of human nature is a tendency to devalue goods that cost little or nothing or appear to have been deeply discounted. Ori Brafman and Rom Brafman, in a recent book titled *Sway: The Irresistible Pull of Irrational Behavior*, describe a 2005 economics experiment in which three groups of students were given a test of mental acuity. The students in the first group were simply given the test. Those in the second group were asked, before taking the test, to drink a bottle of an energy drink called SoBe, which, they were told, is promoted as an intelligence-enhancer. "These students," Brafman and Brafman write, "also had to sign an authorization form allowing the researchers to charge $2.89 to their university account for the SoBe." The students in the third group were also given SoBe and the same explanation about its alleged ability to enhance intelligence but were told they would be charged only eighty-nine cents because the university had received a discount. The results? Those who paid $2.89 for their SoBe scored slightly better on the test than those in the control group—while those who paid the reduced price scored "significantly worse." Brafman and Brafman write, "Given that the SoBe beverage was exactly the same for both groups who received it, we can only conclude that it was the *value* the students attributed to the SoBe that made the difference. . . . Once we attribute a certain value to something, it's very difficult to view it in any other light."[48] Other experiments have shown similar results with variably priced medications and other products and services. The danger in making transit travel free would be that doing so might reduce its value in the eyes of users, possibly contributing to an increase in crime, vandalism, or general neglect. Maintaining even a nominal charge for transit use might reduce that possibility, while also hugely swelling the funds available for transit expansion and improvement. Charging even a dollar for a subway ride—half the 2008 fare—would annually generate a billion dollars that could be spent on additional trains.*

Another questionable element of the Kheel plan is the elevated charge for trucks entering the central business district: thirty-two dollars (charged once per day), versus sixteen dollars for cars. Enemies of urban traffic congestion often focus on trucks, primarily because trucks make life so unpleasant for the drivers of smaller vehicles, both when the trucks are in motion and when they're parked (or, as is often the case in Manhattan, double-parked). Komanoff, when I asked him about the truck charge, told me that it seemed fair to him to charge vehicles in proportion to the "road space" they occupy. Trucks certainly do take up more room than cars do and therefore seem to contribute more to the clogging of streets, but they don't pose anything like the same environmental problem. In fact, in cities, trucks actually possess extremely high environmental utility, analogous to the

---

*The plan raises the possibility of temporarily maintaining some subway charges: "Barrier-free transit promises a new sense of mobility that will make the whole city a resource for everyone to enjoy. If the public is reticent about going the whole way right away, intermediate steps might be to charge a somewhat lower toll and maintain peak period subway fares while offering free local bus trips (which are already heavily subsidized) and free off-peak subway service which is more responsive to price and for which there is ample capacity."[49]

utility of density itself, since without them many of the environmental benefits of population density would not exist. When
Ann and I lived in Manhattan, there were two grocery stores and
a half-dozen specialty-food shops within a block of our apartment, an abundance that made it easy for us to do all our food
shopping on foot—and all of those stores, quite obviously, had
to be supplied by trucks. A transit-and-walking city, by necessity,
is also a truck city, so any policy that makes life harder for truck
drivers also, indirectly, makes life harder for transit riders and
pedestrians. Here in Connecticut, our regular UPS man has to
drive 125 miles to make one day's deliveries; in Manhattan, by
contrast, a UPS driver can often deliver a comparable number of
packages while parked at the curb in front of a single tall building. It's trucks that make efficient, high-density living feasible,
and it's high density, in turn, that makes urban truck use so
efficient. Rather than penalizing trucks—through a fee structure
that treats them, in effect, as twice as bad as cars—any comprehensive transportation plan should explore ways to ease their
access to the city and to its businesses. Trucks in cities are not the
enemies of transit; they help to make it work.

The Kheel plan, nevertheless, contains a number of highly
desirable and urgently needed provisions, among them a proposal to raise the bridge and tunnel tolls for all cars entering
the city, and to extend those tolls to the bridges that currently charge nothing, as well as a provision that would eliminate
some of the traffic lanes that congestion pricing would empty—
a crucial but usually forgotten element of any congestion-

fighting plan. Also sound is a proposal to greatly reduce the number of free curbside parking spaces in Manhattan. There are currently 45,000 such spaces below Ninety-sixth Street, and the Kheel plan would reduce that number to 12,000, by adding 33,000 meters. Free parking is actually an issue that needs to be addressed nationwide. One of the most obvious ways to discourage unnecessary automobile use, while also generating funds to support new and more efficient transit, is to make car drivers pay more of the true cost that their driving imposes on others. Manhattan contains some of the most valuable real estate in the world, yet the city simply gives away, and maintains without charge, many hundreds of acres of it, in the form of free curbside parking spaces.

IN 2008, DAIMLER AG, THE COMPANY THAT MANUFACtures the Mercedes-Benz, introduced American drivers to the Smart Fortwo, a tiny two-passenger automobile that the company has sold in Europe since 1999 and is now marketed in three dozen countries. The Fortwo is only 106 inches long, or nearly three and a half feet shorter than a standard Mini Cooper (and just six feet longer than a Cozy Coupe, the plastic toddler-powered vehicle sold by Little Tikes). In 2007, just before the official launch, I spent part of an afternoon riding around Manhattan in one with Dave Schembri, who is the president of Smart USA, and I had many opportunities to observe public reaction to the car. As we headed up Third Avenue, for example, a passenger in a green Mercury Mountaineer with suitcases bungee-corded to the roof made the international roll-down-your-window signal, then hollered, "I saw you on *Good Morning America*!" and gave Schembri a thumbs-up. This, Schembri said, was a typical response, even from SUV drivers.

At one intersection, Schembri pulled up next to a Toyota Matrix, which had stopped at a light, in order to prove that the Matrix was practically Escalade-size by comparison, and a cop, apparently objecting to the fact that Schembri had driven into the opposing traffic lane in order to do this, stopped his own car alongside the Fortwo and gave Schembri a meaningful look. Schembri shouted, "Hey, we're just test-driving it! It would make a good police car—you'd get to know your prisoner real well!" The cop smiled and drove away. This response, Schembri said, was also typical: during several days in the city he hadn't received a single ticket, despite having parked in some very peculiar places. "The other day, we were in Manhattan with a film crew," he continued, "and four parking enforcement people came up to us and said, 'We would never ticket that car.' It's made for New York."

Is there anything not to like about a car that gets two or three times the city gas mileage of a Hummer? Actually, there is. The world would be better off if people who now drive gigantic gas guzzlers did their driving instead in tiny, fuel-efficient vehicles, but, even so, Fortwos and other micro-cars are not necessarily

environmentally friendly. As with most putatively green innovations, their true impact depends on how they're used. Fortwos
and similar cars, precisely because they are so small and inexpensive, are a potent environmental negative if they make driving a
temptation for people who now get by without cars. You can
easily parallel-park two Fortwos in any curb space that's long
enough to fit a single Lincoln Navigator L—or perpendicular-
park three of them, since the Fortwo is less than a third as wide
as the Navigator is long, and is so short that it doesn't stick out
into traffic when docked nose in. This, I saw repeatedly, makes
the Fortwo deeply appealing to city dwellers who currently think
of car ownership as lunacy—and that's a bad thing. Schembri,
at one point during the time we spent together, perpendicular-
parked our Fortwo in the seemingly too-small gap between two
parallel-parked police cars on the east side of Third Avenue, to
show me that the car could be squeezed into a space that looked
barely big enough for a Harley-Davidson. (This method of parking is actually illegal in Manhattan, although the cops on either
side appeared not to mind.) A Brooklyn resident stopped to
admire the car, asked many questions, and then said, "We need
those in Bay Ridge. Everybody would have a place to park." He
asked the price—around $12,000 for the basic model—and said
he might buy one, since it would enable him, finally, to stop
using the subway. A few months later, *The Wall Street Journal*'s
reviewer found the Fortwo inconceivable as a replacement for
the average American car, but did note that it was extremely easy

to park on crowded city streets—"just fitting the car where others can't is as thrilling as city driving gets," he wrote—and that it had promise as an "urban runabout."[50] That, from an environmental point of view, should be considered a very serious problem, not a solution. The world does not need an inexpensive car that tempts city dwellers to abandon public transit.

Actually, most reviews of the Fortwo have been quite negative. In particular, drivers haven't liked the five-speed automatic transmission, which causes the car to lurch between gears, or the fuel economy, which is in the low 30s much of the time—a remarkably unimpressive performance for a vehicle that weighs just 1,800 pounds and has a 1-liter, 3-cylinder, 70-horsepower engine. "That's worse than my Honda," a passerby complained to one of *The New York Times*'s four reviewers, Eric A. Taub, who test-drove the car in California. It's also not that much better than my Subaru station wagon, if you factor in the additional cost of premium fuel, which the Fortwo requires. Taub concluded, "With its limited carrying capacity, seemingly mediocre fuel economy, erratic handling and fitful acceleration, one question that potential buyers in this part of the world should be asking is, what's the point?"[51] All of the *Times*'s reviewers, however, reported an extraordinary level of public interest. That was certainly my experience. While Schembri and I were talking, yet another pedestrian stopped. "This car is good for China, India, and Europe," he told me. "Especially China. I'm from India." He walked a little way down the sidewalk, so that he could examine it from a different vantage point, and asked, "How many people is it for?"

"Two," Schembri said. "That's why it's called 'Fortwo.' But there's a lot of room. We had a six-foot-eight guy in there, and he fit beautifully. Also a guy who weighed more than five hundred pounds. He said it was the only car he'd been able to get in and out of without a shoehorn." The man from India considered this, then said, "But if you make the front a little shorter, and add a little extra in the back, and you put the trunk under the seat, and maybe make it a little bit longer, you have a five-people car. You know what I'm talking about?"

This, too, is a common line of automotive reasoning: if a small car is a good idea, then a bigger small car must be a better idea. In 2008, BMW introduced a new Mini Cooper, the Clubman, which is ten inches longer than the original. This is partly the result of feature creep, or high-tech reverse entropy, which causes machines and man-made systems to evolve in the direction of greater complexity, and partly the result of the fact that people who love small cars tend to love everything about them except their size.

In 2007, a team of engineers at the Massachusetts Institute of Technology, with technical and financial support from General Motors, designed a tiny concept vehicle they called the City Car, a high-tech electric two-passenger automobile even smaller than the Fortwo. The team's website says that City Cars are intended for "dense urban areas." The vehicles are designed to be compactly "folded" and "stacked," like grocery carts or airport luggage carts, at recharging stations situated near subway stops and other "key points of convergence," causing them to occupy far less curb space than conventional cars. (The front end of one City Car fits into the rear end of any other, so that when several are parked in a line they require less room.) The basic idea is that users will travel by public transit to the stop nearest their destination, and then, rather than walking the rest of the way, will swipe their credit card, board the City Car at the head of the line, and drive. City Cars have many gee-whiz technological features, among them computer screens instead of dashboards, joysticks instead of steering wheels, and "self-contained, digitally-controlled robotic wheels," which can turn 360 degrees, "making it a lot easier to park in really tight urban spaces." William J. Mitchell, who is the director of the Design Laboratory at MIT's Media Lab, where the car was conceived, told an MIT audience in 2007 that the City Car will be perfect for Manhattan. "A city block currently accommodates about 80 vehicles," he said. "With the City Car, that number could be increased to 500." A brochure shows the cars being charged by rooftop solar panels and hydrogen-powered fuel cells. Ryan Chin, a Ph.D. candidate and the project's coordinator, said, "The idea is to have the vehicle work in unison with its urban surroundings, taking advantage of existing infrastructure, such as subway and bus lines. Ultimately we see this as an effective way to merge mass transit with individualized mobility, creating a new urban transportation ecosystem."[52]

But the City Car, as described by its inventors, is a good idea only if you believe that not being able to find a parking space is an environmental problem, and that dense urban areas have something to gain from getting pedestrians off their feet and into cars. Residents of dense urban cores largely get by without individual vehicles now; what would be gained by turning those people into drivers of high-tech golf carts? The fact that City Cars might be recharged by photovoltaic panels is beside the point: wasted energy is wasted energy no matter how it's generated. The City Car is a characteristically complex technological approach to a transportation problem for which urban dwellers long ago devised a vastly superior, low-tech solution. The transportation challenge faced by the residents of New York City and other dense urban centers, no matter what the engineers at MIT say, is not how to fit more cars onto their streets. The City Car's inventors sound exactly like Henry Ford and Frank Lloyd Wright in the 1920s as they enumerate the likely benefits of increasing the use of automobiles.

The idea that big, inefficient cars are an *urban* environmental problem is a good example of the illogic (and the anti-city bias) that runs through much American thinking about the environment, since it is actually in non-dense suburbs and exurbs— places like my town, where traffic jams are not an issue but efficient public transit is essentially inconceivable—that downsized motor vehicles have a necessary role to play. Most of the car trips that suburbanites take obviously don't need to be taken in seven-passenger "light trucks" that get twelve miles to the gallon, and those trips could easily be handled by much smaller, much lighter, much more fuel-efficient vehicles. (The main job performed by any car is moving the car itself. A 120-pound soccer mom, alone behind the wheel of her Toyota Land Cruiser, represents only 2 percent of the total load being moved by her vehicle's gas-burning engine. The other 98 percent is Land Cruiser.) But in suburban use there's no need for any of the fancy features that consumed the creative energies of the MIT engineers: cars in the suburbs don't need to "stack," and a communal curbside rental system serves no purpose in any area where virtually all travel is by automobile and, therefore, drivers have nothing to gain by borrowing cars rather than owning them. The entire City Car concept, including its reliance on an extensive existing transit system, makes sense only in places that don't need it, or shouldn't have it. You have to wonder whether the MIT engineers ever truly asked themselves what problem they were trying to solve. Did they really think that the environment would come out ahead if they could find a way to turn urban pedestrians into drivers?

Four

The Great Outdoors

As population density increases, transit use increases, too, until a certain high level is reached, at which point the graph begins to flatten. This fact is sometimes cited as evidence that the transportation-related environmental benefits of population density are not unlimited, and that there is nothing to be gained by allowing density to rise beyond some relatively low level by (for example) permitting the construction of buildings taller than four stories. But the reason the graph flattens is that, once development becomes sufficiently compact, even trains and buses begin to seem inefficient for many trips, and people simply walk or ride bikes. Reaching that point is a worthy goal, in places where it can be achieved, because, from an environmental point of view, walking and bicycling are the ideal forms of public transportation. The transit systems in New York, in European cities, and in other dense urban areas are relatively energy efficient, in comparison with other forms of motorized transportation, because they receive concentrated use in confined regions, but they don't run on air: transit leverage is not infinite. The New York subway system is powered by electricity, which is generated by burning natural gas, and the trains consume enough of it to light all of Buffalo, a city with a population of almost 300,000.[1] Walkers and cyclists require energy inputs, too, in the form of increased food intake, but their extra eating doesn't entail the burning of as much fossil fuel as the operation of electric trains does, and walkers and cyclists require far less complicated infrastructure. The greatest environmental gains from population density arise once destinations become so close to one another that people elect to get around all by themselves—the urban-transit equivalent of the point at which a nuclear chain reaction becomes self-sustaining.

Despite the obvious environmental benefits of leg power, the only places in the United States where significant numbers of people still use walking as a major form of transportation are New York City, San Francisco, and a few other urban areas where density is high and residential and other uses are thoroughly mixed. In the suburbs you seldom see anyone on foot who is actually traveling to a destination rather than merely moving between a building and a vehicle or trying to lose weight. This is not because city dwellers are virtuous and suburbanites are not. Rather, walking is dependent upon density in the same way that transit is. If you dilute the concentration of people and destinations, walking stops. In the United States as a whole, walking by adults dropped 40 percent between 1977 and 1995.[2] That decline didn't occur because Americans got lazier (even if they did); the gradual extinction of walking is a result of the way we have chosen to arrange the places where we live and work and shop.

In fact, the increasingly sedentary American way of life is more nearly a symptom than a cause of low-density, car-dependent development. The walking destination closest to Ann's and my front door, here in Connecticut, is our mailbox, which is 150 yards down the driveway and up the road. When we lived in New York City, the same 150-yard walk, beginning at the door of our apartment building, could have taken us to any of two dozen plausible destinations, including six or seven restaurants, a shoe-repair shop, a liquor store, two grocery stores, various doctors' offices, a pharmacy, and a half-dozen large apartment buildings, which, taken together, contained the homes of more people than the total population of our current hometown. No wonder we walked more when we lived in the city than we do now that we live in the country: we had many more places nearby to go to.

The sprawl-generated decline in walking may have consequences for health as well as for energy consumption. The life expectancy of a New York City resident is nine months longer than that of the average American, and daily walking may be part of the reason. Clive Thompson, in an article in *New York* magazine in 2007, cited an explanation from Eleanor Simonsick, an epidemiologist at the National Institute on Aging, in Bethesda, Maryland, who, in Thompson's words, had pointed out that New

York is "literally designed to force people to walk, to climb stairs—and to do it quickly."[3] Beginning in 2001, Simonsick and seven other researchers studied a large group of septuagenarians for six years and found that a subject's average walking speed at the start of the study correlated strongly with the likelihood of his or her being alive and in good health at the end of it.[4] Thompson wrote, "As Simonsick sees it, the very structure of the city coerces us to exercise far more than people elsewhere in the U.S., in a way that is strongly correlated with a far-better life expectancy. Every city block doubles as a racewalking track, every subway station, a StairMaster." A British study concluded that every minute spent walking extends life expectancy by three minutes.[5] In 2004, Lawrence Frank, a professor in the School of Community and Regional Planning of the University of British Columbia, in Vancouver, found that white men living in a relatively dense, mixed-use community in Atlanta weighed an average of ten pounds less than a matched group of men living in a typical suburban subdivision in the same metropolitan area.[6]

It's possible that the suburban association between lack of exercise and obesity runs in the other direction, and that overweight people naturally gravitate to cars and sprawling subdivisions while hard-charging ectomorphs are more likely to end up striding the sidewalks of downtown financial districts. (In some parts of the country, cars function partly as devices for transporting air-conditioning between buildings.) But the connection between urban density and regular walking is indisputable. If you want walking to be a significant form of transportation, you have to put people and destinations close together. Willpower alone cannot turn walking into a reasonable means of getting around.

People who live in cities walk more than people who live in suburbs, but not all cities are equal. During a recent visit to Washington, D.C., I asked the concierge of my hotel for walking directions to a certain other hotel, where I had an appointment. He was surprised that I would consider making the trip on foot. "It's a long way," he said. "You should take a cab." I walked anyway, and later determined, by using the ruler feature on Google Maps, that the distance I had covered was just about exactly one mile. No Manhattanite or European city dweller would ever question a healthy person's intention to walk a mile. (It's the distance up Madison Avenue from Fifty-ninth to Seventy-ninth Street.) But the layout of a city like Washington—where buildings are shorter and spaced farther apart, and where businesses and residences are less likely to be mingled, and where the street plan can be disorienting, especially for visitors—makes distances seem longer than they do in a truly dense urban core, where highly varied human activity bustles along the entire route. Genuinely compact development makes distance seem less intimidating and thus increases the likelihood that someone making a short trip will elect to make it on foot rather than in a vehicle. The transportation benefits of density, therefore, depend not just on actual proximity but also, partly, on perception.

This phenomenon can be seen within Manhattan, too, in those parts of the city where the buildings are even more widely separated than they are in the District of Columbia. Few New

Yorkers would think twice about walking west on Forty-second Street from Lexington Avenue, on the east side of Grand Central Terminal, to Times Square, a distance of about three-quarters of a mile. But many would choose not to walk from Fifth Avenue, on the east side of Central Park, to Lincoln Center, on the West Side, even though the distance is the same. The reason is that the first route follows one of the city's liveliest streets, while the second transects Central Park, whose broad empty spaces make all walks seem far longer. You can test this yourself, by observing the pedestrian traffic moving from one side of Central Park to the other. There isn't much, even if you include people who are jogging for exercise. People traveling to a destination are far less likely to walk across a park or any other large open space than they are to walk the same distance along a lively city street.

Jane Jacobs discusses this trick of perception in *The Death and Life of Great American Cities*, in a chapter she called "The Curse of Border Vacuums." She defines a border as "the perimeter of a single massive or stretched-out use of territory" and cites, as the classic example, railroad tracks, whose division of urban areas into isolated, noninteracting regions is so well-understood as to be a cliché: "the wrong side of the tracks." Jacobs writes, "The root trouble with borders, as city neighbors, is that they are apt to form dead ends for most users of city streets. They represent, for most people, most of the time, barriers."[7] Borders deaden urban vitality by breaking the circuit of human activity. They create streetscape voids, which repel pedestrians and make even relatively short distances seem forbidding. And they don't need to look like obvious barriers in order to have that effect. Huge single-use buildings, wide streets, big parking lots, and leafy, spacious parks can be every bit as stifling as railroad tracks. Their effect, furthermore, is cumulative. In a typical suburban commercial strip, the alternation of low, single-use buildings and oversized, single-business parking lots makes walking virtually unthinkable, even if the actual distances between neighboring destinations are relatively short.

Central Park presents an extremely interesting case, because most New Yorkers like it so much in theory that they seldom stop to think about how they use it in fact. As Jacobs points out, "the park's perimeter—especially on its west side—contains great vacuous stretches, and it exerts a bad vacuum effect along a lot of border. Meantime, the park is full of objects, deep inside, that can be used only during daylight hours, not because of their nature but because of their location. They are also hard to reach for many of their putative uses."[8] Large urban parks, including Central Park, have many of the same drawbacks that sprawling suburbs do: they insert so much space between individuals and uses that they actually inhibit many of the activities they are intended to encourage. Central Park covers 843 acres. Those acres would work far better, and function less as a barrier to the overall human flow, if they had been situated somewhere other than in the center of the city, or were penetrated by many more artificial attractions (concession stands, restaurants, sports facilities, museums, playgrounds, theaters) designed to generate and sustain an unbroken chain of lively human interaction all the way across the park, or—perhaps ideally—if the expanse had been broken into smaller, less imposing units and distributed around Manhattan. Central Park contains many appealing features—a children's zoo, two skating rinks, a reservoir surrounded by a running track, an outdoor theater, many playgrounds, tennis courts, a variety of playing fields, the Metropolitan Museum of Art, many others— but the most heavily used features, day in and day out, have always been the ones situated along the park's outermost edges, closest to the surrounding streets, and they gain nothing by being placed on the same large plot of ground or being isolated from one another by banks of vegetation. Central Park is like a mountain range that functions as an impenetrable divide between the valleys on either side, turning them into distinct ecosystems; it is the main source (and main preserver) of the marked cultural differences between the Upper East and Upper West Sides.

Washington Square Park, in Greenwich Village, covers a much smaller area, less than ten acres, but on a square-footage basis it is far more heavily used than Central Park, even though Central Park attracts an estimated 25 million visitors a year. Washington Square Park draws a steady flow of pedestrians into itself because it's fully surrounded by human activity—including New York University, for which it serves as a virtual quadrangle—and is compact enough so that anyone entering it can usually see a human presence extending all the way through to the other side. These are features that keep it occupied, and therefore functioning, late into the night. By contrast, there are sections of Central Park where many New Yorkers (and almost all tourists) feel unsafe on foot, even during the day. This is less true now than it was thirty or forty years ago, thanks mainly to the heroic efforts of New York's Parks Department, but it's still true. And large city parks in more marginal or less dense neighborhoods present greater deterrents to walking from one side to the other. Prospect Park, in Brooklyn, covers 585 acres (and was designed shortly after Central Park, and by the same two men, Frederick Law Olmsted and Calvert Vaux). It's a classic piece of nineteenth-century urban landscaping, and it probably preserves more genuinely natural topographical features than Central Park does, but it's even more resistant to penetration by pedestrians, who are often reluctant to venture more than a short distance beyond the surrounding sidewalks, even in daylight. The broader and more densely vegetated an urban park is, the less safe people feel away from its edges. The isolated central regions of Central Park and Prospect Park act as borders within borders, discouraging park users on either side from walking all the way across, and from taking full advantage of interior features.* And the same is

---

*During some hours of the day, one of the main functions of Central Park and Prospect Park is to provide shortcuts and speedways for taxicabs and other cars—some of whose trips are made necessary by the perceived difficulty of walking from one side to the other. Park users often complain about this traffic, which makes the parks less of a natural refuge and poses a safety threat to bicyclists, skaters, runners, and others attempting to use the same roads. But automobile traffic does introduce a human presence into interior park regions where pedestrians might otherwise be less willing to venture. Cars are obviously not the most desirable form of human presence for a park, but activity does promote activity. Pedestrian use of Central Park would fall if the park were not crisscrossed by streets. Making it more "natural" would cause it to be used less.

true of similar parks in big cities in other parts of the world, among them London's Hyde Park and Kensington Gardens.

Central Park contains many acres of playing fields, and when Ann and I lived in New York I sometimes made use of them. Several friends and I played in pick-up touch football games in the park on summer weekends, and sporadically for a year or two we played in a coed field hockey game on Thursday evenings. But even activities like these would be served better by a different park scheme. Central Park's fields are too concentrated in one part of the city, making it difficult for residents of distant neighborhoods to access them. They are also largely hidden from the view of pedestrians outside the park. Most of the park's playing fields and other large-scale recreational features would have worked better if they had been distributed around the city or if they had been pushed toward Manhattan's outer edges, which, because of the rivers and the expressways that run along them, function as border vacuums anyway. A good example of this sort of intelligent urban design is Riverbank State Park, a twenty-eight-acre recreation area that was built in the early 1990s on top of the North River Wastewater Treatment Plant, a sewage processing facility that extends into the Hudson River on the far Upper West Side. Riverbank contains playing fields, tennis and basketball courts, an indoor Olympic-size swimming pool, a skating rink, a fitness center, a theater, a restaurant, and other attractions, and it is accessible to pedestrians by means of two bridges that cross the Henry Hudson Parkway from Riverside Drive. Riverbank also provides something not usually available to most Manhattanites: a close-up view of the water that surrounds the island. (Manhattan is unusual, among relatively small islands, in that most residents, most of the time, have no sense whatsoever of being surrounded by water.) Riverbank contains no natural landscape features, and is therefore far more obviously artificial than either Central Park or Prospect Park, but it works better, acre for acre, since it attracts and concentrates outdoor recreational activity in a part of the city that wouldn't have attracted it otherwise, and it does so without creating human dead zones at its center and on the streets that border it or lead toward it.

In recent years, the city has invested millions in upgrading and creating innovative recreational areas along Manhattan's outer edge—among them Hudson River Park (a joint venture with the state) and the Manhattan Waterfront Greenway, a well-marked thirty-two-mile route for walkers, runners, skaters, bicyclists, and other nonmotorized travelers which follows the Hudson, Harlem, and East rivers along the island's perimeter, with occasional inland detours (across Dyckman Street, way up beyond the Cloisters; along the spine of central Harlem; around a couple of dozen blocks near the United Nations). The Greenway is especially well suited to bicyclists, who, if they are moderately fit, can cover the entire distance in a single leisurely morning or afternoon. Biking the Greenway turns the city inside out, as I myself discovered on a spectacular Friday afternoon in the fall of 2008. During my circumnavigation, which began and ended in Battery Park, at Manhattan's southern tip, I saw much evidence of the city's successful ongoing effort to reclaim its shoreline as welcoming, accessible public space: tennis courts, batting cages, esplanades, concession stands, kayak-rental centers, picnic areas, art exhibits, and other attractions along a five-mile stretch of the Hudson River, in an area that used to be dominated by garbage, rats, and rotting piers; a cluster of basketball half-courts under the West Side Highway; various volunteers tending the public gardens in Riverside Park; a man fishing with an enormous surf-casting rod a little downstream from the George Washington Bridge; a beautifully renovated boathouse, and a trailer carrying rowing shells, in Swindler Cove Park, on the eastern edge of Harlem; a group of schoolchildren receiving roller-hockey instruction on a playground in Carl Schurz Park, along the East River; more bollards, cleats, capstans, hoists, and other riverside mooring paraphernalia, some of it freshly painted, than you would think could possibly have an ongoing nautical application in New York City; and many people playing soccer, touch football, and tennis in a beautifully landscaped, mile-long strip park at the edge of the East River, on either side of the Williamsburg Bridge. All of these recreational spaces are heavily used, and, because they are arranged along the city's outer edge in areas that would otherwise discourage pedestrian flows, they are green in a way that a monumental nineteenth-century creation like Central Park can probably never truly be. They enhance urban vitality without undercutting the features that make population density both environmentally benign and culturally appealing, and for that reason they are extremely valuable models for twentieth-century urban planners.

City streets can function as border vacuums, too. Park Avenue above Forty-fifth Street—four lanes, divided by a grassy median—divides the Upper East Side into two mostly non-overlapping pedestrian zones: the six-block-wide area between Park Avenue and the East River, and the two-block-wide strip between Park Avenue and Central Park. Park Avenue is much narrower than Central Park, but it bifurcates the East Side almost as starkly. Pedestrian movement is heavy on both sides of the divide in the blocks that lead up to Park Avenue, but it seldom crosses over—an effect compounded by the fact that no buses operate on Park along the 3.75 miles between Grand Central Terminal and 120th Street. In that sense, this long stretch of roadway closely resembles the broad boulevards of central Washington, D.C., since it serves the convenience of motorists to the detriment of pedestrians and transit users, and since its streetscape consists mainly of single-use buildings (apartments, hotels, office towers) with virtually none of the street-level commercial hubbub that makes walking in other parts of the city appealing, productive, and self-reinforcing. One of the worst offenders in this regard is the building that architects probably celebrate the most, Ludwig Mies van der Rohe's Seagram Building, on the east side of Park between Fifty-second and Fifty-third streets. That building is surrounded by a broad plaza, a feature that has been much admired and imitated but is actually very lightly used (except by smokers) and constitutes a dead zone within a dead zone—a border vacuum appended to a border vacuum. The Seagram plaza, perversely, inspired the city to amend its zoning regulations in 1961 to encourage other developers and architects to isolate their own buildings in the same way. This form of separation is one of the oldest ideas in zoning: the belief that pushing buildings farther apart adds human value. A fad for buildings surrounded by empty plazas followed, with a corresponding decrease in true pedestrian-friendliness—a good example of how unreliable New Yorkers can be at identifying what they truly like about their own city, and of their occasional tenacity in undermining the very qualities that make Manhattan efficient and user-friendly.

Several other features of Manhattan's general layout work against pedestrians as well, most notably the long crosstown distances, above Fourteenth Street, between the avenues on the West Side. Midtown pedestrians can sometimes overcome this difficulty by using fortuitous shortcuts, such as building lobbies that extend all the way from one cross street to another, but the long blocks make diagonal travel more difficult than it would be if the blocks were shorter. Suburban subdivision plans also often thwart walkers (and bicyclists) when, as is often the case, they feature winding lanes that either loop back on themselves or suddenly terminate. As John Holtzclaw as written, "The traditional rectilinear street grid, with short blocks, offers many alternative paths, allowing the walker to explore different streets, find favorites, and to link trips more easily. Winding streets, intersected by dead ends and cul-de-sacs, require longer trips and allow no such variety."[9] Gridlike street plans may seem unimaginative, but they increase the mobility of pedestrians and are almost self-explanatory: walking in much of Manhattan, even in areas where the distance between avenues is great, is like walking on a map.

THE OPTICAL ILLUSION THAT MAKES A PEDESTRIAN LESS likely to walk three-quarters of a mile across a park than to walk the same distance along a busy street is even more pronounced in non-dense, nonurban settings. During the seven years that Ann and I lived in Manhattan, I'm fairly certain, I never took a cab just eight blocks, no matter what the weather, yet I have often driven from my house to the little post office on our village green—exactly the same distance, four-tenths of a mile. There are only a few houses along my route to the post office, and that lack of density makes the walk seem discouragingly long, even though I know that this impression is just a trick of perception. Our daughter, Laura, was born in Manhattan (in a hospital that we walked to from our apartment when Ann went into labor), and when she was very young her favorite activities included being taken for walks in her Snugli. If she was crying at night, a quick hike up and down Second Avenue would almost always either cheer her up or put her to sleep. When she was a little older and could sit up in a backpack, she and I would sometimes accompany Ann on her morning walk to work, in midtown—a round trip of more than three miles. My intention was always to turn around immediately if Laura began to seem bored or unhappy, but she seldom complained, and we not only got all the way to Ann's building most times but, often, made significant detours on our way home, doing errands en route, with no sign of unhappiness from her. When we moved out of the city, in October 1985, shortly after Laura's first birthday, I was excited to think about how much more she would enjoy going for walks in the country, which at the time of our move was at the peak of New England's leaf season. But she didn't love it at all. The first time we walked to the village green to buy the morning newspaper, on a spectacular autumn morning, she fussed and squirmed in her backpack almost the whole way. As far as she was concerned, there was nothing to look at. The absence of urban commotion along our route made the walk seem long and boring. And it usually has the same effect on me, although I hate to admit it.

Laura's and my experience contradicts a notion held by many environmentalists, who maintain that nonnatural landscapes discourage people from going outside at all. Douglas Farr, in a recent book called *Sustainable Urbanism: Urban Design with Nature*, writes, "The unpleasant characteristics of today's outdoor spaces are especially harmful in close urban settings, actually deterring people from spending time outdoors and reinforcing the tendency to stay indoors and close the windows."[10] There's no denying that the air I breathe in northwestern Connecticut seems cleaner to me than the air I breathed in midtown Manhattan, but Farr is wrong about what keeps people inside. In fact, the only parts of the United States in which large numbers of adults with indoor jobs regularly spend significant periods of time outdoors are "close urban settings," and that's because those are the only parts of the United States where walking remains a practical and necessary form of getting around. A resident of a dense city almost can't help logging at least an hour or two outside every day, just doing things like walking to work, walking to lunch, walking to the subway, and walking to perform various errands. For a typical adult who lives in a nonurban residential area, by contrast, routinely spending a comparable amount of time outdoors requires a powerful external motivation, such as ownership of a riding lawnmower, a New Year's resolution to finally get some exercise, or (as in my case) an addiction to playing golf. Going outside is actually a more normal, ordinary activity in a dense city, because there it's an indivisible element of daily life. (The average visitor to the Grand Canyon, by contrast, spends something like ten minutes actually under the sky. The standard drill: arrive by bus, look over the edge, go to the bathroom, visit the souvenir shop, buy something to eat and drink, return to the bus.) Every winter, here in Connecticut, there are stretches of several days when I spend virtually no time at all outdoors, except when I'm taking the dogs out to pee in the snow, or walking to and from the garage. That was never true when we lived in New York.

A couple I know made this same discovery in reverse. They recently moved back to Manhattan part-time after living in the country for twenty years, and discovered that being surrounded by open spaces in Connecticut had caused them to fall almost entirely out of the habit of going anywhere on foot. In fact, the wife was so out of the habit that at first she found walking more than a block or two to be physically uncomfortable. But walking in dense city is nearly as unavoidable as driving is in suburbia, and she now walks to work, to the grocery store, to restaurants, everywhere. Both husband and wife have found that they get far more exercise now, as city dwellers, than they did when they lived full-time just a step from the great outdoors—and they also feel the sense of liberation that comes from genuine mobility. This is true for children, too. If you want to see large numbers of kids fooling around outdoors, away from school, don't walk through a suburban residential neighborhood, where it's unlikely you will see anyone of any age.

Dense cities can be very good places for older people, too. Here in Connecticut, I play bridge with a number of women in their seventies and eighties. They all dread the moment, which is not far off for some of them, when they will have to give up their cars, because in non-dense residential areas the loss of a driver's license represents an almost total end to independence, a virtual house arrest. In a dense city, by contrast, a driver's license is not a necessity, and older people have almost all the same transportation options that younger people do. Routine walking, furthermore, keeps them fitter and thus extends their ability to get around without help from others. In New York, Ann and I occasionally visited an elderly widow who had an apartment in our building. She lived well into her eighties, and because none of her regular activities required her to own a car, she remained fully independent until shortly before she died.

Environmentalists and urban planners sometimes say that, in order to get people out of their cars and onto their feet, developed areas need to incorporate extended "greenways" and other attractive, vegetated pedestrian corridors. It's true that such features, along with parks and natural areas, can encourage some people to *take walks*. But if the goal is to get people to embrace walking as a form of practical transportation, oversized greenways can actually be counterproductive. Walking-as-transportation requires closely spaced, accessible destinations, not broad expanses of leafy scenery. In my town's principal commercial district—a picturesque collection of small businesses along the edge of a river—people will very often drive from the grocery store to the bookstore on the other side of the main street, even though the two buildings are separated by less than a hundred and fifty yards. The reason is that there are only two intermediate destinations between those two businesses—two small offices of the same real estate company—plus a vacant lot, and even at the micro-scale of a small town, this lack of density can make a walk seem forbiddingly long. And virtually nobody walks from the grocery store to the dry cleaner, just another fifty yards farther away.

Residents of my town, when making a trip to the grocery store, often complain that there is nowhere to park if the dozen spaces directly in front of the store's entrance happen to be occupied when they arrive, even though there are almost always dozens of empty spaces a short walk away. These people, in fact, will sometimes circle the parking area, waiting for a space to clear, rather than park at the other end of the lot, fifty yards farther on.* The same trick of perception is to blame. When these same people go to the nearest shopping mall—which is a forty-five-minute drive from our town and has a parking lot that is several times the size of our entire business district—they think nothing of parking much farther from the mall's entrance than they would ever think of walking between stores at home.

Of course, once they are inside the mall they also walk from store to store without complaint, and cover far more territory on foot than they would ever think to do while running errands in our little village center at home. This is not a paradox. The distances inside the mall don't seem as long as the distances at home do—and the reason, again, is density. A shopping mall, for all its considerable flaws, embodies some of the main organizational principles of Manhattan and other dense cities: the

---

*Environmentalists often fret about the useless miles that urban drivers log while circling city blocks in search of empty parking spaces, and they have proposed many supposedly green remedies, including the installation of electronic sensors to alert drivers to the location of vacant spots. But the likely difficulty of quickly finding a parking space is one of the major factors that make urban drivers reluctant to drive, and is therefore good for the environment, no matter how much pointless circling undeterred drivers end up doing. Schemes intended to make parking spaces easier to find have the same encouraging effect on driving that adding more parking spaces does: they're just a method of inducing additional traffic. My town has a popular restaurant whose parking lot is too small to easily handle peak weekend crowds. This fact is well known, and, as a result, couples dining together sometimes double up— "Parking will be tight, so we'll pick you up"—something that wouldn't happen if parking were unlimited. You can't reduce the environmental damage of cars by making parking easier.

stores are close together, with no mandated buffers or setbacks between them, and they front directly on wide pedestrian walkways. The layout of a mall's interior, in other words, is designed to meet the needs of walkers rather than of drivers—even though malls themselves, quite obviously, are entirely car-dependent and therefore contribute to the sprawl that makes them necessary.

NEW YORKERS DON'T NECESSARILY APPRECIATE THE REAL reasons that walking is such an important element of their daily life. In the late 1990s, Rudolph Giuliani, then the mayor, undertook a campaign to eradicate jaywalking—a major issue for him. Pedestrian barriers were erected near a number of midtown corners, stoplights and pedestrian crosswalks were shifted away from some intersections, and the police were instructed to issue summonses to pedestrians who crossed streets mid-block or against a light. The policy was almost universally ignored, by cops as well as pedestrians, and it was widely ridiculed. ("Yes, many of us who live in New York City did think Mayor Rudolph Giuliani was joking when he said he was cracking down on jaywalking," Calvin Trillin wrote in *Time* in 1998. "But then a law student crossing Sixth Avenue got a $50 jaywalking ticket. What we had forgotten was that Mayor Giuliani is never joking."[11]) The policy was also thoroughly misguided. In Manhattan, creative jaywalking is an environmental positive, because it makes traveling on foot easier: it enables pedestrians to maintain their forward progress when traffic lights are against them, and to gain small navigational advantages by weaving between cars on clogged side streets—and it also keeps drivers on their guard, forcing them to slow down. The real purpose of anti-jaywalking laws is not to protect pedestrians but to make life easier for drivers. That's why anti-jaywalking rules are enforced (and observed) in Los Angeles, where the cars are entirely in charge. Rather than banning jaywalking, cities should take steps to enhance and enforce the rights of pedestrians, and to impede cars in areas where traveling on foot is feasible. (One useful step would be to follow New York City's good example and make it illegal for drivers to turn right on red lights.) Tightly controlling pedestrians with a view to improving the flow of car traffic just results in more and faster driving, and that makes life even harder and more dangerous for people on foot or on bikes. In fact, studies have shown that pedestrians are safer in urban areas where jaywalking is common than they are in urban areas where it is forbidden.

A concept that has been very popular among forward-thinking urban planners in recent years is that of "traffic calming," which is the practice of designing and managing roads in ways that force the vehicles that use them to operate at lower, safer speeds. The most common traffic-calming tool in the United States is probably the speed bump, but there are many more sophisticated ones: the elevation and clear marking of crosswalks (to underscore the priority of pedestrians), the mixing of road shapes and textures (to keep drivers attentive), the expansion of curbside parking, especially angle parking (to narrow the travel lanes, thereby impeding drivers and forcing them to slow down), the addition of marked bike lanes (to reduce the space for cars and make cyclists a more conspicuous presence), the planting of tall trees near curbs (to "reduce the optical width" of roads, in the words of Peter Newman and Jeffrey Kenworthy), the use of four-way stop signs (which reduce intersection accidents by large percentages in comparison with both two-way stop signs and traffic signals), and the complete removal of selected travel lanes (to constrict vehicle flow and make the reclaimed real estate available for other uses, such as widened sidewalks).[12] In 2008, I traveled to the United Kingdom with several American friends who had never visited that country. One of my friends was amazed—and appalled—that in small village centers drivers were often allowed to park on both sides of the already narrow main roads, leaving less than two full travel lanes between them. This seemed irrational to my friend, since the constricted road space forced the drivers moving in both directions to slow down repeatedly, and to take turns passing through the narrowest places. But these impediments, whether entirely planned or not, had the distinctly beneficial effect of keeping in-town traffic speeds at safe levels and of simplifying the lives of pedestrians, who could fearlessly cross more or less at will.

A very closely related modern traffic-calming concept is known as "shared space," which is a technique for controlling traffic by blurring (rather than sharply delineating) the boundaries between driving areas and walking areas; by making strategic use of traffic-impeding "street furniture," such as plantings, benches, and bicycle racks; and by eliminating traffic lights, stop signs, lane markings, and other traditional controls. This sounds to many people like a formula for disaster, but the clear experience in the (mainly) European cities that have tried it has been that increasing the ambiguity of urban road spaces actually lowers car speeds, reduces accident rates, and improves the lives of pedestrians: drivers proceed more warily when they aren't completely certain what's going on. (Shared space is also actually an ancient idea, since it's pretty much the way all large public areas functioned before the rise of automobiles.) The Dutch have a very similar traffic-calming tool, the *woonerf*, or living yard, which is a road that is designed intentionally to keep drivers guessing, by blending the spaces used by cars and pedestrians, and by placing thought-provoking obstacles in the paths of drivers. I once experienced much the same thing on a large dock in Massachusetts. Cars and trucks were allowed on the dock, along with a great deal of other human activity, yet there were no painted lines, signs, or other clear indications to show where anything was supposed to go. As a result, everyone moved slowly and cautiously, and took extra care to stay out of everyone else's way. Similarly, in small-town centers the installation of the first traffic light is typically followed by a rise in accident rates, as drivers begin to rely more on signals than on sense.

Streets in Manhattan have actually employed primitive (and mostly unintentional) versions of all these calming techniques for decades. Potholes, double-parked FedEx trucks, jaywalkers, dog walkers, crosswalks jammed with pedestrians and baby carriages, protruding Dumpsters, football-throwing teenagers, ututilty workers smoking cigarettes near barricaded manholes, high-rise construction equipment, building scaffolding, horse-mounted traffic cops, bicycle messengers, and gridlocked intersections all function as traffic-calming instruments, and they long ago turned many Manhattan streets into "shared spaces," to the continuing benefit of pedestrians. The fact that obstacles like these often infuriate drivers, mainly because they force them to slow down, is a good thing for pedestrians, transit users, and the environment, no matter what Rudy Giuliani thinks. Manhattan would not be a better place for anyone if sidewalks were enclosed behind fences, and cars on side streets were routinely able to travel at forty or fifty miles an hour. Urban congestion-pricing schemes—such as those discussed in the previous chapter—always carry the danger that, by making car traffic move more smoothly, they will eliminate the (mostly accidental) features that make urban walking efficient and appealing.

When cities set out to give special treatment to pedestrians, they are often as misguided as when they attempt to thwart them. The most common such step is the permanent or temporary conversion of regular city streets into vehicle-free zones—in effect, the creation of outdoor shopping malls. Sometimes this works, but often it's counterproductive. Stores and other businesses can't exist without vehicles to serve them, and the rather specialized physics of pedestrianism doesn't automatically cause walking to expand to fill any space that is provided for it. Christopher Alexander, Sara Ishikawa, and Murray Silverstein—architects who were associated with the Institute of Environmental Structure at the University of California at Berkeley—wrote, in their influential 1977 book, *A Pattern Language*, "It is common planning practice to separate pedestrians and cars. This makes pedestrian areas more human and safer. However, this practice fails to take account of the fact that cars and pedestrians also need each other: and that, in fact, a great deal of urban life occurs at just the point where these two systems meet. Many of the greatest places in cities, Piccadilly Circus, Times Square, the Champs-Élysées, are alive because they are at places where pedestrians and vehicles meet. New towns like Cumbernauld, in Scotland, where there is total separation between the two, seldom have the same sort of liveliness."[13] In addition, if sidewalks are wide enough, most pedestrians will continue to use them, in preference to blockaded streets, unless the streets are densely filled with mini-destinations of their own. Urban places that are accessible only to pedestrians can thrive, but they need to be properly scaled and situated, and enlarging them doesn't necessarily improve them. Enormous public spaces are like oversized living rooms: people tend to avoid them, or to keep to their edges. Anyone who has ever given a large party has observed this phenomenon indoors: guests accumulate first in the smaller spaces—the front hall, the kitchen, the laundry room—and colonize cavernous rooms only tentatively, when the cozier nooks are overflowing. You can see exactly the same tendency in large public squares. Huge empty spaces, both indoors and outdoors, resist human occupation.

In New York in 2008, Mayor Bloomberg introduced a program called Summer Streets, which banned motor vehicles from about seven miles of Manhattan streets, including Park Avenue below Seventy-second Street, on three Saturdays in August between seven in the morning and one in the afternoon. The plan was greeted enthusiastically by most residents, environmentalists, and bicyclists (although some expressed concern that the elimination of all motor traffic on some streets would merely increase congestion on others). Paul Steely White, the executive director of the advocacy group Transportation Alternatives, told *The New York Times*, "It's a new way to use a street, using it more as a park than as a thoroughfare. Everyone around the world knows about Park Avenue as one of New York City's most storied thoroughfares, and to turn that over to pedestrians and cyclists, even though it's just for three consecutive Saturdays, I think that sends a very powerful message that the tide is turning so that bicyclists and pedestrians are on at least an equal footing with drivers."[14] There was nothing actually wrong with the plan, if one considered it as a special recreational event, like allowing kids to play around fire hydrants on hot days, or turning a large parking lot over to a carnival. (One of my happiest memories of New York is of a winter day—it was the late seventies or early eighties—when a huge snowstorm shut down virtually all vehicular traffic in the city, and people cross-country skied on Third Avenue, around the corner from our apartment.) But programs like Summer Streets don't really lead anywhere, in terms of broad transportation strategy for urban areas. City bicyclists undoubtedly enjoyed being able to tool down Manhattan's spine, virtually unimpeded, from the middle of Central Park to the Brooklyn Bridge, but removing motor traffic from upper Park Avenue didn't make that street any less of a pedestrian wasteland than it already was. Car-free programs like Summer Streets treat pedestrians and bicyclists the way Robert Moses used to treat cars, by segregating them on expressways of their own.

A better idea, which Bloomberg's office announced a few weeks later, was a plan to reconfigure Broadway by closing two of its four lanes to vehicle traffic and turning those lanes over to pedestrians, bicyclists, and vendors. That program actually carried New York a step closer to something that every big city needs, which is a surface-transportation vision that integrates buses, trucks, cabs, cars, bicyclists, and pedestrians into a coherent system in which all elements coexist safely, while steadily shrinking the space devoted to cars.

ONE OF THE MAIN REASONS THAT PEOPLE MOVE TO THE suburbs is to acquire ground of their own, yet most suburbanites end up discovering that, all things considered, they don't like being outside or allowing their children to be outside—a reluctance reflected in, and sustained by, our increasingly common anxieties about skin cancer, West Nile virus, Lyme disease, allergies, poison ivy, pet-eating coyotes, rabid raccoons, car traffic, sexual predators, kidnappers, terrorists, and other perils associated with allowing your front door to close behind you. A friend once told me that the airplane passenger sitting next to him on a business trip to Los Angeles offered him a dollar for every swimmer, of any age, that he spotted in any of the hundreds of backyard pools that came into view as they approached LAX. My friend saw none, and the man said he had made the same offer on other flights and had never had to pay off. This phenomenon is not confined to Los Angeles, or to swimming pools. You can travel for miles through suburbia and see no one doing anything in a yard other than working on the yard itself—often with the help of a riding lawnmower, one of the few four-wheeled passenger vehicles that get worse gas mileage than a Hummer. In 2006, a researcher at the University of Montana, in a study based on satellite data collected by the National Aeronautics and Space Administration, determined that the nation's largest irrigated crop is cultivated grass, which covers more than 32 million acres in the continental United States. (The second largest irrigated crop, at roughly 10 million acres, is corn.)[15] Homeowners spend more than $40 billion a year on their lawns, and they use approximately a hundred million pounds of pesticides, which they apply more heavily than farmers do. A third of all residential water use, furthermore, goes into yards.[16] The modern suburban yard is perfectly, and perversely, self-justifying: its purpose is to be taken care of.

Beyond caring for their lawns, the interest of most Americans in almost any direct experience of the outdoors is falling precipitously. Bicycle riding is down by nearly a third since 1995.[17] A 2008 study published by the National Academy of Sciences showed that participation by Americans in a variety of outdoor activities had declined by as much as 25 percent since the late 1980s, as measured by the frequency of visits to national parks, the number of applications for hunting and fishing licenses, and other indicators. Patricia A. Zaradic, who is a conservation ecologist and a coauthor of the study, said, "Folks are going out into nature much less and decreasingly every year. It would take 80 million more visits this year to get the per capita number back up to the level it was in 1987."[18] This reversed an earlier trend, which the Sierra Club had described, in 1963, as "the extraordinary upsurge of hiking and camping and boating and the overwhelming increase in use of our national parks," which had begun following the Second World War.[19] The decline has accelerated in recent years; between 1995 and 2005, overnight camping in national parks fell by 24 percent.[20] These changes have occurred even as the National Park System has made itself accessible and appealing, with features and amenities aimed at nearly everyone on the broad mobility spectrum from wheelchair users to backcountry trekkers.

In 2006, Zaradic and the other coauthor, Oliver R. W. Pergams, of the University of Illinois at Chicago, published a related paper, in which they described the same growing tendency to stay indoors, and coined a word for it, "videophilia." "Increased use of video games, home movies, theatre attendance and internet combined with inflation adjusted oil prices explains the majority of the 16-year decline in per capita US national park visits," they wrote. "The average person in the US went from spending 0 h/year on the internet in 1987 to spending 174 h/year on the internet in 2003, and from spending 0 h/year playing video games in 1987 to spending 90 h/year in 2003. . . . Watching movies at home and in theatres increased another 63 h/year during this period. Altogether, the average person in the US spent 327 *more* hours/year on these entertainment media in 2003 than he or she did in 1987, an incredible increase in time."[21] Pergams and Zaradic found a similar trend in Japan, and in a 2008 interview with a representative of the Nature Conservancy, which funded both their studies, they said, "It seems clear that major industrialized countries are moving towards videophilia. As the most videophilic societies, the United States along with Japan may be the first to experience its consequences—but it's likely that other nations are not far behind."[22] They wrote that they suspected that decreasing interest in outdoor recreation would further weaken Americans' already fragile interest in the natural world and that the effect might be especially strong in children, who (as Pergams and Zaradic wrote in a 2007 paper, also funded by the Nature Conservancy) "reportedly spend an average of 30 minutes of unstructured time outdoors each week" and whose relationship with nature in adulthood is shaped by childhood experiences.[23]

Even when you do see kids outdoors nowadays, they often don't seem fully committed to the concept. A few years ago, I drove past a baseball game on a school playing field and noticed that the left fielder was talking on a cell phone. One of the trends that Pergams and Zaradic have tracked in their research is a steady decline in camping—but even for the dwindling die-hards camping as it's done nowadays is an awful lot like staying at home. In 2006, my wife and I and our two kids rented an RV in Las Vegas and spent two weeks visiting national parks in northern Arizona and southern Utah, and even though we our-selves were in a gas-guzzling, air-conditioned box we felt that we were practically roughing it in comparison with most of our neighbors at various campgrounds. We saw televisions and satel-lite antennas and fancy propane grills, and the RV parked next to ours at a campground near Bryce Canyon was pulling an enormous trailer loaded with minibikes, dirt bikes, and four-wheeled all-terrain vehicles, and almost anywhere we went we had to keep an eye out for seven-year-olds on miniature motor-cycles. Even in the true outdoors, interaction between children and the environment nowadays is often mediated by internal combustion engines. There were large stretches in southern Utah in which we noticed that almost any unfenced terrain within a reasonable distance of the highway was covered with ATV tracks.

ATV drivers at least have to leave their houses. In more sub-urbanized parts of the country, people seem increasingly unin-terested in being outside at all. An article by June Fletcher in *The Wall Street Journal* in 2007 described growing homeowner dis-satisfaction with upscale "outdoor rooms"—trendier, more ex-pensively furnished versions of the open-air spaces that used to be known as porches, patios, and decks—which had been popular suburban upgrades during the previous decade but had apparently lost their appeal as their owners had experienced what being outside was actually like. "The backyard misery has been a boon for exterminators and repair shops," Fletcher wrote. "Fire ants nest in speakers and televisions. (They're attracted to the hum and vibration.) Squirrels chew on the arms of teak furniture and on speaker wires. When expensive electronics come into contact with water, dust, pollen and heat, burnouts and other problems can occur. Over the past two years, such issues have boosted service requests at Walt's TV & Home Theater in Tempe, Ariz., by 400%." Fletcher cited an academic study of twenty-four middle-class households in Los Angeles which found that even when backyards "were equipped with pools, patios, grills and, in one case, a skateboard ramp, children spent little time playing in them and adults rarely used them. More than half of the families spent only 'negligible' amounts of time in their yards, mostly doing chores."[24]

The study Fletcher mentioned was written by Jeanne E. Arnold, a professor of anthropology at the University of California at Los Angeles, and Ursula A. Lang, an architect in Berkeley. They based their observations on data collected by UCLA's Center on Everyday Lives of Families. "In quite a few of these cases," they wrote, "no family member so much as stepped into the back yard. Sporadic activities in other cases were confined to non-leisure chores such as taking out trash or briefly feeding dogs or washing off chairs." The authors found that even though the homeowners had made "extensive invest-ments" in their outdoor spaces, they nevertheless "largely admire them from afar—from inside the house or in their mind's eye while busy doing other things." Even so, the subject families' backyards, driveways, and garages were often cluttered with toys, furniture, and other unused recreational items, leading the authors to conclude that the storage of material goods had become "an overwhelming burden for most middle-class fami-lies, especially in the West, where basements are generally not able to absorb possessions." Most of the families in the study, the authors noted, no longer parked cars in their garages, which they had converted into storage spaces for bikes, toys, outdoor furniture, athletic equipment, and other possessions they hardly ever used, "with the resigned understanding that storage is going to be the sole long-term use." This suggests that, even in a culture heavily shaped by cars, cars themselves come second to stuff.[25]

I'm ashamed to admit that Ann and I fall into this same category, despite not living in the West. A decade ago, our insur-ance company ordered us to tear down our garage, which had appeared to be on the verge of collapsing ever since we bought our house, a dozen years before. The structure was actually stur-dier than it looked, but we complied. This created a problem not for our cars, which we usually kept in the driveway, anyway, but for the things we had stored in our garage because we couldn't think of what else to do with them: all our bikes, sleds, skis, hedge clippers, firewood, and snow shovels; a big broken plastic swimming pool filled with small broken plastic things; our significant collection of trash cans; two charcoal grills; our old lawnmower, which we no longer need, since we have gotten tired of pushing it around and have hired a landscaping service to cut our lawn; some spare tires; an unused croquet set; a huge pile of old broken bricks; and a large squatter community of squirrels. With few exceptions, these are possessions we couldn't have owned in Manhattan, because we wouldn't have had places to put them: the more space you have, the more stuff you fill it with. A friend of mine owns a self-storage facility in our town, and its many units have been fully rented virtually since the day he opened for business. He says that he's amazed at what people pay to store there: useless items they couldn't resist acquiring and now can't bear to unload. His facility is landfill purgatory for the indecisive. I'm not a customer, but when Ann and I built a new garage we made sure it had a large upper story. The new garage contains more enclosed space than our old Manhattan apartment did, yet we use most of the space for no purpose more productive than archiving junk we vaguely hope our kids may someday want for their kids, plus several old broken bicycles.

In 2005, Richard Louv, who attended the same terrific Colorado summer camp that I did, back in the 1960s, wrote a book called *Last Child in the Woods* in which he lamented the decline of outdoor playing and identified an increasingly common childhood ailment, which he called "nature-deficit disorder."[26] In 2007, he addressed the same subject in a magazine article titled "Leave No Child Inside." He wrote:

Within the space of a few decades, the way children understand and experience their neighborhoods and the natural world has changed radically. Even as children and teenagers become more aware of global threats to the environment, their physical contact, their intimacy with nature, is fading. As one suburban fifth grader put it to me, in what has become the signature epigram of the children-and-nature movement: "I like to play indoors better 'cause that's where all the electrical outlets are." His desire is not at all uncommon. In a typical week, only 6 percent of children ages nine to thirteen play outside on their own.[27]

Louv's childhood—like mine—was very different from this. He used to surreptitiously remove survey stakes from vacant lots in his neighborhood, "to slow the bulldozers that were taking out my woods to make way for a new subdivision." My own friends and I knew this same dread (although we also knew the joy of playing in half-built houses and ransacking them for plywood scraps and other tree-house-building supplies, as I'm sure Louv did also). We eyed the few vacant lots in our neighborhood with the same gnawing anxiety that members of the Nature Conservancy feel as they watch untouched woodlands being clear-cut to make way for strip malls. I vividly remember standing in my backyard with a group of friends, at the age of nine or ten, and watching in horror as a backhoe hired by the new owner of the house next door ripped out a dense tangle of brush at the end of his backyard which we had treated for years as our own jungle.

But this point of view is deeply hypocritical. The ultimate source of the survey stakes that Louv and I and our friends dreaded and despised was the very same yearning for personal turf that had pushed our parents to buy or build our own houses, which were right next door. Louv and I and other kids of our generation were no different from the subdivision-building intruders; we just got to those neighborhoods a few years sooner. Louv, when reminiscing about his childhood, portrays himself as a victim of sprawl, but his family arrived in his childhood neighborhood on the same human tide that, when the next wave crested, obliterated the woods next to his house. And the sense of impending loss that Louv felt as a child is the same force that causes people like him and me to yearn to live in places the bulldozers haven't reached yet, never noticing or acknowledging that by doing so we are drawing the bulldozers right behind us. We all tend to think of ourselves as the last unsinning inhabitants of whatever place we live in. We don't usually recognize ourselves as participants in its destruction.

A sensitive person's first reaction to the mounting evidence that Americans, especially young Americans, may be losing interest in directly experiencing the natural world is likely to be one of regret and loss, or even despair. But is it necessarily a bad thing, globally speaking? It seems perverse to say so, but sitting indoors playing video games is easier on the environment than any number of (formerly) popular outdoor recreational activities, including most of the ones that the most committed environmentalists tend to favor for themselves. In the end, it may not be a bad thing for the earth or for the human race if increasing numbers of Americans would rather watch our shrunken wilderness on TV than fly to it in an airplane and drive across it on a motorbike. Do you need to have climbed Mount Everest yourself in order to value the Himalayas? (The vast garbage dumps that cover the most heavily traveled lower reaches of Everest were created by climbers, not by people who hate mountains.) The main cultural force behind the steep rise in National Park attendance in the decades following the Second World War was not a sudden new interest in wilderness conservation but the broadening ownership of cars, which, for most people, made long-distance travel affordable for the first time. If the changing economics of driving forces us to drive less, we will spend less time in the places our cars used to carry us to. Will that necessarily be a bad thing for the earth, or even for ourselves? American environmentalists have traditionally assigned little value to human activities that take place away from thoroughly natural settings. But that is an attitude that, in the end, is unsustainable.

Environmentalists have tended to think of themselves mainly as defenders of what's left, rather than as shapers of what lies ahead. Their focus has been on erecting and maintaining barricades around as much unspoiled territory as possible, and, in the manner of Jefferson and Thoreau, on denigrating what they view as ways of life antithetical to that ideal. Protecting undefiled natural areas is an unassailable goal, but Jefferson and Thoreau are poor role models for a world with a rapidly growing supply of people and rapidly shrinking supplies of fuel, clean water, and other resources. The future of humanity will be predominantly urban. From an environmental point of view, we need to apply ourselves to making city life appealing and life-enhancing, not to wishing that doing so were unnecessary. As Richard Louv and I and millions of other American children of our generation discovered in our youth, pulling up the survey stakes next door doesn't work.

On a radio program I heard in 2008, an environmentalist who was being interviewed advocated the reactivation of the Civilian Conservation Corps, an extremely popular Depression-era federal relief program that put unemployed young men to work on a broad variety of conservation-related maintenance and construction projects all over the country, primarily in state and national parks. The program ended officially in 1942, but it lives on through dozens of similar state and federal programs, most of which also employ young people and many of which descended directly from the original. The environmentalist on the radio argued that reviving the CCC would both perform a useful environmental service and create jobs at a time of high unemployment. Those are both worthy goals, but from a truly long-term perspective it might be more useful to turn the problem inside out, by putting the same people to work improving the quality of human life in dense urban centers—a Civilian City Corps—with a view to making compact urban living more attractive and appealing to a larger proportion of the population, thereby creating an environmentally beneficial counterforce in opposition to sprawl.

Five

Embodied Efficiency

n 2003, the National Building Museum, in Washington, D.C., held a show called *Big & Green: Toward Sustainable Architecture in the 21st Century.* A book of the same name was published in conjunction with the show,[1] and on the book's dust jacket was a photograph of 4 Times Square, also known as the Condé Nast Building, a forty-eight-story glass-and-steel tower in New York City between Forty-second and Forty-third streets, a few blocks west of Grand Central Terminal. (The offices of *The New Yorker,* for which I am a staff writer, occupy two floors in the building.)

When 4 Times Square was completed, in 1999, it was considered by many to be a major breakthrough in urban construction. Daniel Kaplan, a principal of Fox & Fowle Architects, the firm that designed it,* wrote in an article in *Environmental Design & Construction* in 1997, "When thinking of green architecture, one usually associates smaller scale," and cited as an example the headquarters of the Rocky Mountain Institute, a nonprofit environmental research and consulting firm based in Snowmass, Colorado, whose staff contributed to the design of 4 Times Square. "It is a tribute to the efforts of RMI," Kaplan continued, "that we can begin to apply green principles to large-scale commercial developments, and it is vitally important to do so."[2]

This idea—that creating an environmentally responsible large building in a dense urban location requires exceptional effort and ingenuity—was widely held at the time and, in fact, is widely held today. The RMI headquarters building, which Kaplan presented as a model, comes much closer to the popular conception of green design. It's a 4,000-square-foot, superinsulated structure with curving sixteen-inch-thick walls and krypton-filled windows, and it's set into a breathtaking hillside at the end of a gravel road in the mountains about fifteen miles northwest of Aspen. Much of its heating is generated passively by an attached greenhouse, which is backed up on very cold days by a pair of woodstoves and, on occasion—as RMI's chief executive officer, Amory Lovins, has said—by the 50-watt body heat of a large dog. ("On really cold nights we'd adjust her to a

---

*Fox & Fowle's name partners were Robert Fox and Bruce Fowle. They later split up. Fox's current firm is Cook+Fox Architects; Fowle's is FXFOWLE Architects.

100-watt dog by throwing a ball."[3]) Virtually all of the building's electricity comes from photovoltaic solar panels, and excess current is stored in a bank of Chinese submarine batteries or sold to the local power company by feeding it into the grid. The building was erected in the early 1990s and serves partly as a showcase for green construction technology. It is also Lovins's home.[4]

Thanks partly to Lovins and RMI, 4 Times Square includes many innovative features, too, among them collection chutes for recyclable materials, photovoltaic panels incorporated into parts of its skin, natural-gas-fired absorption chillers that provide heating and cooling, and curtain-wall construction with exceptional shading and insulating properties. In terms of the building's true ecological impact, though, these and other overtly green innovations are distinctly secondary. (The photovoltaic panels, as is often the case in high-profile green construction projects, are really just for show; they supply less than 1 percent of the building's electric power, and usually supply none at all.) The two greenest features of 4 Times Square are ones that RMI and most other people never even mention: it is big, and it is in Manhattan.

Environmentalists have long characterized large urban buildings as intrinsically wasteful, primarily because they represent high levels of "embodied energy"—energy that was consumed in the fabrication and transportation of the materials from which they were made, and that was expended in the process of construction—and because so much interior space has to be given over to elevator shafts and other mechanical systems, and because such buildings place intensely localized stresses on sewers, power grids, and water systems. The show *Big & Green* and the book associated with it were put together by David Gissen, who is the curator of architecture and design at the National Building Museum, and in his introduction he takes this conventional point of view by suggesting that creating "environmentally sensitive" big buildings requires quite a stretch, since big buildings "consume enormous amounts of energy, release large amounts of carbon dioxide, use the most wasteful construction techniques, and have poor air quality that can cause numerous illnesses."[5] Nowhere in his essay does Gissen mention the most truly significant environmental fact about big urban buildings, which is that density creates the same kinds of energy-and-emissions *benefits* in individual structures that it does in entire communities. Tall multistory buildings, whether or not their designers intended them to be green, have much less exposed exterior surface per square foot of interior space than broader, lower buildings do, and that means that they present relatively less of themselves to the elements, and that their compact roofs absorb less heat from the sun during cooling season and radiate less heat from inside during heating season, no matter what they're made of. A study by Michael Phillips and Robert Gnaizda, published in *CoEvolution Quarterly* in 1980, more than two decades before the opening of *Big & Green*, found that an ordinary apartment in a typical building near downtown San Francisco used 80 percent less heating fuel than a new tract house in Davis, a little more than seventy miles away, and used less energy in all categories, even though the tract houses had been built in accordance with a new building code that emphasized energy conservation.[6] This beneficial effect is amplified in dense urban cores, where one building often directly abuts another. A magazine editor I know who grew up in a row house in Philadelphia in the 1950s told me that his parents always knew when a house on either side of theirs had become vacant, because their heating bills would spike. "Sharing walls shares and saves heat," John Holtzclaw has written. "Exposing less wall and roof area to the sun reduces summer air-conditioning loads."*[7]

Tall buildings, furthermore, help to create the concentrations of people and uses which are necessary to sustain far greater environmental benefits, such as efficient transit systems and compact networks of civic services, and to eliminate the reckless waste created by the helter-skelter duplication of freeways, schools, fire departments, power stations, postal delivery routes, sewage-treatment facilities, and innumerable other high-cost, high-energy public amenities. Although the elevator shafts required by tall buildings fill significant amounts of interior

---

*During the summer, this energy benefit can be counteracted to varying degrees by the so-called urban heat-island effect. Cities are usually significantly warmer than rural areas in the same climate zones, mainly because buildings and paved areas absorb and then radiate solar energy, as well as emitting concentrated amounts of waste heat from buildings and vehicles. The excess heat increases warm-weather cooling loads, decreases cold-weather heating loads, affects outdoor comfort levels, and alters precipitation patterns.

space, elevators, because they are counterweighted and thus require less motor horsepower, are among the most energy-efficient passenger vehicles in the world: moving people vertically through a city requires less energy and less infrastructure than moving them horizontally. (Cable cars, which were introduced in San Francisco in 1873, work on a similar principle. The cars going downhill help to pull the cars going uphill, like weights on either end of a rope running over a pulley.) Stacking and concentrating dwellings and businesses is the easiest way to make communities truly efficient, and it is the only way to achieve deep reductions in per-capita energy use and carbon output in large, prosperous populations, not least of all because residents of urban apartment buildings are less likely to own and use cars. (The apartment dwellers in Phillips and Gnaizda's study drove only a quarter as many miles as the residents of the single-family houses.) All of these benefits, along with many others, could and should be thought of as elements of what might be called urban buildings' "embodied efficiency"— which is a far more accurate and revealing measure of any building's overall environmental impact than its "embodied energy" is, since it looks beyond the moment of construction to the building's use over its lifetime.

In 2004, on a reporting assignment, I spent two nights at the Renaissance Scottsdale, a hotel in Paradise Valley, Arizona, one of the many indistinguishable strip-mall-dominated sections of the rapidly expanding Phoenix metropolitan area. The hotel is a paragon of embodied *in*efficiency. It has 136 rooms and thirty-five suites, all contained in single-story "casitas," which are spread over what the website describes as "25 acres of lush land-scaped beauty." The grounds could actually be characterized more accurately as a series of interconnected asphalt parking lots fringed by nonnative vegetation, pockets of overwatered lawn, and a few sorry-looking acres of bulldozed desert. The casitas, which are individual structures and therefore require individual heating and air-conditioning systems, represent close to the maximum possible level of energy inefficiency, since in each one the ratio of interior floor area to sun-baked flat roof is 1:1. My appointment the next day was in a shopping mall directly across the street. I decided to walk rather than drive my rental car but soon regretted my decision because I had trouble finding an easy way out of my end of the hotel grounds and then had to scale a couple of roadside barriers and cross six lanes of suburban traffic and the mall's own large parking lot, which—as is usually the case—covered more ground than the buildings it served.

An instructive counterexample is provided by another hotel in the same chain, Renaissance New York, at the northern end of Times Square, a few blocks from the Condé Nast Building. It has more than twice as many rooms as its Arizona cousin—305 of them, plus five suites—yet its total real-estate footprint is only about a quarter-acre, or just 1 percent the size of the site of the of Renaissance Scottsdale. (The rooms in the New York hotel are divided among twenty-six floors, a far more energy-efficient configuration, and the building extends right to the property line and is bordered by sidewalks rather than by thirsty invasive plantings.) Most guests at the Renaissance New York arrive without cars, since the hotel is a short walk from one of the largest public-transportation hubs in the world and is surrounded by theaters, restaurants, stores, and other attractions. At the Renaissance Scottsdale, by contrast, I saw guests driving from their rooms to the main desk. (I walked to dinner and breakfast, but my route from my casita, which was near the back of the property, took me across a quarter-mile of hot macadam, and I saw no one else on foot.) The Renaissance Scottsdale has two pools, four tennis courts, and a putting green—outdoor amenities undreamt of in Times Square—but during the two days I spent there I saw no one using any of them, although I did see one person typing on a laptop near one of the pools. Most of the other guests who were on the property were doing what people tend to do at hotels all over the world, no matter where those hotels are situated or how many putatively lush acres their grounds cover: they were attending business meetings, eating, sleeping, or watching TV in their rooms, and those who had left the grounds had invariably done so in cars, since, realistically, there were no local attractions that were accessible on foot or by public transit.

EMBODIED EFFICIENCY AND INEFFICIENCY ARE EASY to depict two-dimensionally. Most American suburbs are arranged a little like the dots on the line in the (kindergarten-caliber) graph that follows, in which the round dots represent residences and the square dots, isolated in single-use zones on either side, represent retail shops and businesses:

If you imagine that your own house is the star at the very center of the line and assume that an inch or so in either direction represents the maximum distance that you and other members of your family would routinely be willing to travel on foot, you can see that, unless you are borrowing sugar from one of your nearest neighbors, almost any trip you take away from your house, including any trip to any store, will require you to use one of your family's cars. Those cars aren't part of the physical structure of your house, but your use of them is every bit as much a part of your home's carbon footprint and overall environmental impact as your incandescent lightbulbs, your furnace, your central air conditioner, and your swimming-pool heater. The number of miles you drive every day is directly determined by where you live in relation to where you work and shop and perform the rest of your life's activities, and those car miles should therefore be considered an indivisible part of the environmental profile of your home, and, therefore, one of the principle elements of its embodied inefficiency. So should the creation and maintenance of the infrastructure network that enables you to live where you do—the roads and schools and stores and hospitals and all the rest. A truly dense city with mixed residential and commercial uses, by contrast, looks more like this:

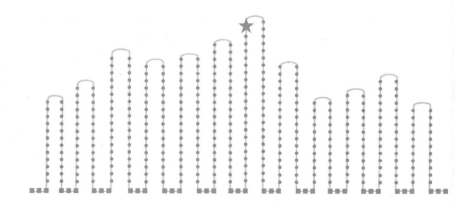

If you imagine that the loops represent tall buildings and that your own apartment is the star near the top of the tallest loop, you can easily see that the same walkable inch, extending in either direction from the base of your building, will bring you, on foot, within reach of many dozens of likely destinations—and that your neighborhood is sufficiently dense that any distance too great to walk can easily be served by efficient public transit. (Actually, Manhattan and other concentrated urban cores are so dense that my little drawing doesn't contain anywhere near enough dots or loops to make the scale remotely comparable to that of my first drawing—but you get the idea.)

No building is an isolated environmental phenomenon, even if a hundred percent of the electricity it uses is generated by a windmill on its roof. Every human structure is just a single element in a large, interrelated energy-and-emissions network, and its impact on the environment goes far beyond the number of BTUs that it consumes directly or that were consumed during the fabrication of the materials from which it was built. The architect Bruce Fowle, whose firm designed 4 Times Square, told me, "The Condé Nast Building contains 1.6 million square feet of floor space, and it sits on one acre of land. If you divided it into forty-eight one-story suburban office buildings, each averaging 33,000 square feet, and spread those one-story buildings around the countryside, and then added parking and some green space around each one, you'd end up consuming at least a hundred and fifty acres of land, and because of all the roofs you would have forty-eight times as much thermal exposure. And then you'd have to provide infrastructure, the highways and everything else." Like many other buildings in Manhattan, 4 Times Square doesn't even have a parking lot, because the vast majority of the 6,000 people who work inside it have no use for one: more than 95 percent of them commute by public transit or on foot. There is a large, delightful cafeteria on the building's fourth floor which is never listed among the building's greenest features but is actually an important one because many Condé Nast employees purchase at least one meal a day there, and when they do they come and go on elevators. What any company's workers do with the waste paper they generate at their desks is far less significant, environmentally, than how they get to and from work and where they go for lunch.

The headquarters of Sprint Nextel Corporation, in Overland Park, Kansas, a suburb of Kansas City, was completed in the early 2000s and is a typical modern corporate "campus." It comes much closer than 4 Times Square or any other Manhattan office building to the popular conception of green design, partly because more than half the site, which covers about two hundred acres, has been preserved as undeveloped land. The campus, which includes office space for approximately 15,000 employees, is touted by Sprint as environmentally sensitive, and it has earned environmental recognition from the federal government and the U.S. Green Building Council, among others. But there is nothing truly green about it. The 15,000 employees are spread among more than twenty low-rise buildings, which together contain 4 million square feet of office space, and because almost all those employees commute by car the campus also has fifteen multistory parking garages, a large surface parking lot, and a subterranean VIP lot, which together provide a little more than one parking space per worker—4 million square feet of parking altogether, or one square foot of parking for every square foot of office space. The complex contains so many separate structures that when the campus was being designed and built the city government of Overland Park had to hire five new planners just to keep up with the inspections and approvals.[8]

Before the campus was completed, a Sprint vice president explained that the idea of constructing a single tall building instead of three dozen scattered small ones had been rejected "because communication doesn't happen as well when offices are vertical instead of horizontal"[9]—a notion that isn't supported either by common sense or by the company's subsequent experience. A local reporter, writing in 2002, observed that "employee interaction took place almost exclusively in the campus' fourteen parking garages, where workers struggled through Sprint's own gridlock each morning and evening," and he quoted a former employee's description of the office buildings' working spaces as "miles and miles of identical cubes."[10] Driving through the campus is an eerie experience. Even by day, the compound looks more like a semi-defunct state hospital or a medium-security prison than an incubator of intellectual exchange, and the few pedestrians you see at most hours tend to be heading either toward or away from one of the parking garages, which ring the central part of the compound.

Even more important, from an environmental perspective, is the fact that the campus was built on former farmland many miles from the city's center (and almost forty miles from the airport, which is about as far to the north of downtown Kansas City as Sprint headquarters is to the south), and has therefore helped to enlarge what was already one of the most disastrously sprawling metropolitan regions in the United States. (According to the Sierra Club, which in 1998 named Kansas City the fifth most "sprawl-threatened" large city in the United States, "Kansas City has more freeway lane-miles per car than any other city in the country."[11]) The cars parked in all those garages, along with the new auto-dependent subdivisions in which many of their owners live and the spreading infrastructure network on which they depend, are the true indicators of the campus's environmental impact, and are the main components of its actual embodied-energy load—its extraordinarily high level of embodied inefficiency. Yet these same negative features have usually been treated—by the company, by civic leaders, by the press—as tremendous positives. In 1998, Shirley Christian, of *The New York Times*, wrote, "Sprint's move is expected to give a huge boost to the already nonstop commercial and real estate development in the area surrounding the campus for several miles. [A spokesman for the general contractor] said at least five hotels are scheduled for construction or in the planning stages near the campus. Residential realtors said that many homes in nearby subdivisions, where most of the houses are 3,000 square feet and up, are being sold to Sprint employees. The upscale Town Center Plaza shopping area, with 73 shops, restaurants and theaters in 750,000 square feet, opened last year across the street from the eastern boundary of the Sprint site."[12] This enumeration of local benefits is really just a sprawl checklist—a catalogue of the

Sprint-driven environmental damage (and added taxpayer and utility-customer expense) that radiates outward from the campus through southern Kansas City and beyond. That damage was magnified in 2005, when Sprint (following several years of corporate turmoil) merged with Nextel, and the combined company moved its corporate headquarters to Reston, Virginia, where Nextel had been based, while leaving most of its employees in Kansas. That move, which required many executives to make weekly round-trips halfway across the country, was reversed in early 2008, when the company decided to move most of its headquarters staff back to Kansas, after all, and to shrink its Virginia operation from nine buildings to seven.

There is nothing truly ecologically enlightened about Sprint's Overland Park campus, no matter how many so-called green features the individual buildings include, or how much recaptured runoff water the grounds staff uses when irrigating the complex's dozens of acres of lawn, or how many bicycles are (theoretically) made available for trips between buildings, or how many trees the company has planted, or how many PETA-certified border collies are used to keep migratory birds from defecating around the edges of the man-made "wetland area" in the southeast corner of the property. The campus is a sprawl bomb, and the "open space" preserved on the property merely makes its impact worse.

Misconceived, ego-driven corporate construction projects like the Sprint complex misappropriate the word "campus," because other campuses, those of non-commuter colleges and universities, have always been among the greenest, most energy-efficient living-and-working configurations in the United States. In fact, residential colleges and universities provide terrific examples of the environmental advantages of increasing population density and mixing uses, Manhattan-style: most residential-college students live in small spaces in energy-efficient, multiunit buildings, and they work and entertain themselves within short distances of the places where they sleep, and they travel among almost all their daily activities by walking, riding bikes, or using public transportation, and they eat most of their meals in centralized dining halls, which reduce meal-related energy use, food waste, and infrastructure duplication. The students' professors tend to live nearby, too, and are therefore also less reliant on cars.

All of these structural features of traditional college life have far more environmental value than most of the high-profile steps that competitive colleges are increasingly taking in the hope of attracting ecology-minded applicants, such as placing solar panels on dormitory roofs, hiring "sustainability coordinators," encouraging students to shorten their showers and do their laundry in cold water, and adding locally grown produce to dining-hall salad bars.[13] The greenest colleges, in other words, are not necessarily the ones that earn the highest scores in the Princeton Review's annual ranking of environmentally friendly campuses; the greenest colleges are the ones whose students are the most likely to live on or near campus and not drive cars. That means that almost any residential college in a dense urban core (Manhattan, Boston, San Francisco) is inherently greener than any college whose campus includes a high proportion of student drivers or is situated far from a public transit system, no matter how large its compost heaps are—although almost any college is still far greener than almost any residential suburb. (The most immediately effective way for a typical residential-college student to further reduce his or her already modest carbon footprint would probably be to stop traveling home for Thanksgiving and to Fort Lauderdale or Acapulco for spring break.) A typical residential college concentrates uses, hugely reducing the environmental impact of the individuals who live and work there, and it therefore has a very high level of embodied efficiency. A typical corporate campus, by contrast, contributes to the separation of uses and acts as a catalyst for cascading environmental damage, and is therefore inherently wasteful and destructive.

RMI's mountainside headquarters building, in most ways, sets an even worse environmental example than the Sprint campus does. It was built in a fragile location, on virgin land 7,100 feet above sea level, and it can hold only a handful of RMI's full-time Snowmass employees, the rest of whom work in a converted (and not particularly energy-efficient) ranch house a half-mile away. Because the two buildings are in a thinly populated area, they force most employees to drive many miles—including trips between the two buildings—and they necessitate extra fuel consumption by delivery trucks, snowplows, maintenance crews, and others, and by Amory Lovins himself when he leaves Snowmass to fly to his numerous consulting and speaking engagements all over the world. Here's a simple graph representing the embodied inefficiency of RMI headquarters:

The star in the center is RMI's main building, Lovins's house. The square dot an inch to the right is the organization's other Snowmass office, where most of the employees work. The round dots represent the handful of other residences nearby, some of which are the homes of RMI employees. The rest of the world lies well beyond the edges of the page. If RMI's employees worked on a single floor of a big building in downtown Denver and lived in apartments nearby, many of them would be able to give up their cars, and the thousands of visitors who drive to Snowmass each year to learn about environmentally responsible construction could travel by public transit instead. And Lovins could ride to the airport on a bus.

Singling out RMI—one of the most respected environmental organizations in the world—may seem unfair, but RMI, along with many other respected environmental organizations, shares responsibility for perpetuating the powerful anti-city bias of American environmentalism, and for discouraging the kind of development that actually reduces community-wide energy use and carbon emissions. That bias leads people to view dense cities as environmental crises, and allows them to believe that the way to shrink their own environmental impact is to follow Lovins's example and head for the hills. It also prevents them from recognizing that RMI's headquarters—which sits on an isolated parcel more than 180 miles from the nearest significant public transit system—is sprawl.

IN 1998, THE U.S. GREEN BUILDING COUNCIL, A NON-profit trade organization, introduced a building-certification program called Leadership in Energy and Environmental Design, or LEED, which was intended to encourage developers, architects, builders, and others to adopt environmentally responsible building practices. LEED has evolved through several iterations since then, and it has grown to include a number of different versions, including ones for new buildings, existing buildings, groups of buildings, commercial interiors, schools, and private houses.

LEED has had a significant impact on parts of the construction industry and on popular thinking about buildings, both for good and for ill. On the positive side, LEED has raised awareness of the environmental implications of building in general, and has helped to spread public awareness and acceptance of various green construction practices, with the result that, in a few high-end markets in the United States and elsewhere, erecting a major building without at least announcing an intention to seek some level of LEED certification is now virtually unthinkable. (There are four rating levels in the LEED system: certified, silver, gold, and platinum.) It has also prompted the upgrading of building codes in parts of the country, has increased awareness of the possibility of recycling many kinds of demolition and construction waste, and has helped to raise manufacturing standards for building components. On the negative side, LEED is expensive and cumbersome to implement—for a large project, the monitoring and certification process alone can consume many hours of consulting time and cost hundreds of thousands of dollars—and it has encouraged the widespread public perception that emission-reduction and energy efficiency are premium add-ons, achievable only with high-priced technology and large teams of advisers, and therefore beyond the reach of ordinary people. LEED is also mainly concerned with individual building features, and has historically given little recognition to how buildings truly function in the communities of which they are a part. Under LEED 2.2, which was the version that was in effect through 2008, for example, a Manhattan office building, such as 4 Times Square, would have received one "Alternative Transportation" credit for being "within 1/2 mile of an existing, or planned and funded, commuter rail, light rail or subway station," while a one-story office building in an industrial park on an interstate highway at the outer edge of metropolitan Phoenix could have earned exactly the same amount of "Alternative Transportation" credit simply by installing bicycle racks and a cyclist changing room, even if every one of its employees actually commuted by car. LEED 3.0, which went into effect in 2009, increases the rail-transit credit to six points. This acknowledgment by LEED's authors of the environmental significance of public transit is a promising step, but it's a decade overdue and it's still deeply flawed because it makes no distinction between buildings whose occupants actually commute by train and those whose occupants theoretically could but don't. It therefore grossly understates the environmental value of siting buildings in high-transit urban cores, like Manhattan and downtown San Francisco, and grossly overstates the environmental value of siting buildings in places like downtown Atlanta, where most office-tower commuters use cars even if they work within hypothetical walking distance of one of the central city's few MARTA stations. There isn't a building in Manhattan that doesn't deserve full LEED certification on the basis of its occupants' use of "Alternative Transportation" alone, and there are thousands and thousands of buildings elsewhere in the country that shouldn't be considered for LEED, or any other form of environmental recognition, because of the energy inefficiency that is unavoidably embodied in the automobile-dependent locations in which their owners decided to place them. Sprint received LEED certification for one of the main buildings on its environmentally disastrous Kansas campus—a disgrace, a joke, or both.

One problem with LEED is that certification is both too easy and too hard. It's too hard because verification requires such a major investment in time, money, and outside consultants. In 2005, Auden Schendler, a LEED-accredited professional and the director of environmental affairs at the Aspen Skiing Company, and Randy Udall, a Colorado environmentalist, wrote an influential article, "LEED Is Broken; Let's Fix It," which drew on their own experiences in numerous construction projects. "We're concerned," they wrote, "that LEED has become costly, slow, brutal, confusing, and unwieldy, a death mark for applicants administered by a soviet-style bureaucracy that makes green building more difficult than it needs to be, yet has everyone genuflecting at the door to prove their credentials. The result: mediocre 'green' buildings where certification, not environmental responsibility, is the primary goal; a few super-high level eco-structures built by ultra-motivated (and wealthy) owners that stand like the Taj Mahal as beacons of impossibility; an explosion of LEED-accredited architects and engineers chasing lots of money but designing few buildings; and a discouraged cadre of professionals who want to build green, but can't afford to certify their buildings."[14]

The LEED-certification process is meticulous in the extreme. The rating criteria are complicated, and the evaluations require third-party verification by high-priced professionals. This is useful if the goal is to prevent builders from receiving recognition they don't deserve, but it's counterproductive if the goal is to achieve the widest possible adoption of more enlightened building practices. One telling fact about LEED is that, as of mid-2008, the program had certified fewer than 1,500 projects in the United States, despite having been in existence for ten years. (The Sprint building was one of just eight certified projects in the entire state of Kansas.) The pace of certification has quickened significantly since 2005, when Schendler and Udall published their critique—at that time, fewer than two hundred U.S. buildings had been LEED-certified, a result that they characterized as "sorely disappointing"—but the program is still focused very narrowly. Schendler and Udall compared LEED's complexity unfavorably with the relatively straightforward simplicity of the federal government's Energy Star program, which was introduced by the Environmental Protection Agency in 1992 and which has transformed the market for all kinds of power-consuming devices in the United States and a number of other countries, including Japan and Australia, as any consumer who has shopped for a household appliance in the past fifteen years probably knows. The U.S. Energy Star program was extended to residential and commercial buildings in 1995, and the EPA expects that by 2010 more than 2 million new U.S. houses will have received Energy Star labels. The EPA also estimates that, in 2007 alone, Energy Star saved Americans $16 billion in direct energy costs and reduced greenhouse-gas emissions by an amount equivalent to those produced by 25 million cars.[15]

One of LEED's shortcomings has to do with its function as a

marketing tool for developers—a source of pride for the U.S. Green Building Council. The council has made much of a 2008 study which showed that office space in LEED-certified buildings rents for $11.24 per square foot more than comparable space in non-LEED buildings, has occupancy rates that are 3.8 percent higher, and sells for $177.00 more per square foot.[16] These are very large premiums, but they are less a reflection of the lower operating costs of LEED buildings than of the cachet that LEED now lends to construction projects in certain upper-level real estate markets. That cachet can encourage developers to pile on high-visibility, low-return features—such as nonfunctioning photovoltaic panels, economically unjustifiable fuel cells, and expensive computer-controlled lawn-watering systems—while ignoring simpler, lower-cost measures that are either less conspicuous or less rewarded by LEED. A critical article about LEED in *Fast Company* in 2007 quoted David White, a climate engineer, who said, "Unfortunately, the exuberant creative stuff—the expensive buzz words such as 'geothermal,' 'photovoltaic,' 'double façade,' and 'absorption chiller'—only makes sense when the basic requirements, such as a well-insulated, airtight façade with good solar control are satisfied."[17] In addition, developers often decide that seeking LEED certification isn't worth the trouble and expense, and that LEED-related fees would be better spent on doing things that actually increase embodied efficiency, regardless of whether or not those features would qualify for LEED credits. (This is a good outcome for the environment, but it undermines the program's rationale.) The new New York Times Company headquarters building, which is fifty-two stories tall and is a short walk from 4 Times Square, is, by virtue of its size and location alone, a vastly greener building than almost any LEED-certified building outside of Manhattan. (It also has many advanced energy-saving features, including an innovative subfloor ventilation system.) Yet the Times Company and its architects—Renzo Piano and FXFOWLE—decided that pursuing certification wouldn't be worth the expense or the aggravation.

Paradoxically, earning LEED certification is sometimes too easy, because the system emphasizes high-tech add-ons and pays little attention to environmental issues that extend beyond building skins and property lines. The first building to be certified LEED platinum, the system's highest rating, was the Philip Merrill Environmental Center of the Chesapeake Bay Foundation (CBF), a nonprofit environmental organization that is dedicated to protecting and restoring Chesapeake Bay and its extensive watershed. The Merrill Center, which opened in 2001, is a two-story, 32,000-square-foot office building situated at the edge of Chesapeake Bay in suburban Annapolis, Maryland. The building's most distinctive exterior feature is its south-facing wall, which is covered by a slotted wooden framework (made partly from recycled pickle-barrel staves) that is intended to prevent excessive amounts of summer sun from shining through the building's windows and raising its interior temperature during cooling season. Somehow, though, none of the architects, engineers, or LEED consultants who were involved in the project noticed that the wooden framework also prevents the sun from shining fully on the building's photovoltaic panels, which were installed behind it. As a result, the panels are crippled most of the time; by the foundation's estimate, they provide only about a half-percent of the electricity used by the building, or about $250 worth per year at 2007 prices. The center has many other conspicuous features of the type that earn LEED credits and make environmentalists and reporters swoon—geothermal wells, composting toilets, cork flooring, bamboo stairs, rainwater-collection cisterns, showers for bicyclists, and a computer-controlled energy-management system that "alerts employees when windows should be opened"—but it has many alarming features, too, beginning with its placement just a hundred feet from Chesapeake Bay, on a thirty-two-acre site ten miles from central Annapolis and an hour's drive from either Baltimore or Washington, D.C., where many of its visitors originate. (Organized one- and two-hour tours of the building are held regularly. The facility can also be rented for parties and weddings.)[18] One of the LEED credits that the Merrill Center earned was for "maximizing open space" by building on only a small part of its large building lot—a credit category that might be thought of as LEED's sprawl reward, since it gives developers an incentive to follow the example of Renaissance Scottsdale by creating low-density projects on open land far from urban cores. "The lot was going to be sold regardless of if we bought it or not," a CBF spokesman said at the time, and he added that the new building was erected on the footprint of an existing one. These are reasonable points. But the spokesman's rationalization is the one that drives all sprawl, and there is nothing green—and certainly nothing platinum—about it. The foundation's previous headquarters was in downtown Annapolis. The new location is inherently less green, since moving away from the central city turned all of the foundation's eighty employees into automobile commuters. The CBF has stated that one of its goals in building the Merrill Center was to set an environmental example for others, but the most significant example it sets is the same misleading one that Sprint and RMI do, because the project suggests that the way to be environmentally responsible is to find an attractive lot in a non-dense location and cover the floors with cork instead of carpet.

Similarly, the Green Globes building rating system—a LEED competitor, which is operated in the United States by an association called the Green Building Initiative—awards ten points to projects that leave at least 10 percent of their building site undisturbed, and another ten points to projects that integrate native plantings into the landscaping on their site—credits that are unavailable to projects in high-density urban cores which, because they fully cover their sites, have no landscaping at all. Green Globes does award ten points for what it defines as dense development, but its density threshold is just 60,000 square feet of building space per acre, the equivalent of a one-and-a-half-story building on a site the size of 4 Times Square's. Green Globes awards up to fifteen points to buildings that use various water-saving measures in the creation and irrigation of their surrounding landscaping, but none of those points are available to buildings that, like 4 Times Square, have nothing to water, even though using no water for landscaping is clearly better for the environment than using less. Green Globes gives points for using high-albedo roofing materials, which are designed to reflect sunlight rather than absorb it, but the award doesn't vary with the size of the roof or with the ratio of the roof's area to the building's interior floor space. A huge one-story building such as a big-box retail store, therefore, would receive the credit if it had a reflective roof, but a tall building with a low-cost, low-tech roof would receive no credit even though its roof, because of its small relative surface area and its thermal isolation from all but a small percentage of the building's interior, would actually represent a greater energy savings. Green Globes also gives credit to buildings that are sited within three miles of an "alternative fuel refueling station"—a LEED-like absurdity.[19]

Schendler and Udall, in their 2005 article, identified two common green-construction mind-sets, which they called "LEED brain" ("what happens when the potential PR benefits of certification begin driving the design process") and "point mongering" ("what happens when the design team becomes obsessively focused on getting credits, regardless of whether they add environmental value"). Both of these mind-sets have become increasingly common, as growing numbers of developers have begun to view at least announcing an intention to seek

LEED certification as a virtual requirement for big construction projects. "LEED can be a way to facilitate regulatory approvals, appease the public, and get free press," Schendler and Udall continued. "There's also a powerful incentive for mechanical engineers, architects and contractors to gain LEED expertise. It labels their firm as 'green,' increasingly a prerequisite on requests for proposals. Unfortunately, if you know how to scam LEED points, you can get the PR benefits without doing much of anything for the environment."[20]

LEED has a fundamental weakness, which is that it is not a comprehensive, objective assessment of true environmental impact but, rather, a values-laden incentive system that encourages projects which adhere to a very particular view of the environment and, especially, to a very particular view of high-end real estate development. The environmental philosophy that underlies LEED has evolved in some important respects, a change that is reflected in the increased value that the rating system began to assign to public transit in 2009. But because LEED is a "portfolio tool"—as I heard it described at an environmental conference intended mainly for developers, architects, and contractors—it tends to favor adding features over subtracting them (because adding features is the economic basis of the building industry), and it tends to favor high-cost, complex solutions over common sense: computer-controlled shading systems rather than hand-operated awnings or venetian blinds; integrated wind turbines rather than smaller windows; rooftop fuel cells rather than connection to an existing off-site co-generation plant. LEED encourages exactly the type of thinking that, in residential construction and remodeling, helped to make grotesquely oversized and over-equipped kitchens and bathrooms the standard even for non-luxury houses. Kitchens and bathrooms are critical rooms for homebuilders because they are the easiest parts of a house to make more profitable. (Piling costly features onto those rooms is the homebuilding equivalent of the funeral-industry practice known as "loading the casket.") LEED has given fancy green-building systems and accessories an economic utility, for builders and manufacturers, something comparable to granite kitchen countertops and freight-car-size refrigerators. The environment doesn't necessarily come out ahead.

Just as important, the building-centric bias of LEED—and, indeed, of almost all rating systems and almost all thinking about the environmental impact of construction—discourages developers, architects, and others from considering projects in any but the narrowest sense. A good example of this way of thinking—a version of the mentality that Schendler and Udall call LEED brain—is a 195,000-square-foot office building constructed by Gap Inc. in San Bruno, California, at the southern edge of metropolitan San Francisco. That building, which was completed in 1997, actually predates LEED by several years, but it has been celebrated around the world as an icon of green architecture, and it's a regular tour stop for architects who want to learn to be more eco-friendly. It has also helped to establish its architect,

William McDonough, as a green guru. (*Time* picked him as one of several dozen "Heroes for the Planet" in 1999, and a profile in *Vanity Fair* in 2008 described him as "a prophet of the sustainability and clean-technology movements."[21]) The building is loaded with ostensibly green features—low-volatility coatings, biodegradable upholstery fabrics, under-floor ventilation, benches made of salvaged eucalyptus wood, soaring, sun-filled atriums, a one-and-a-half-acre undulating roof covered with six inches of soil and planted with native grasses—but the truly critical environmental facts about the building are that it's two stories tall and is served by a multilevel parking garage, and that it's directly accessible only by freeway, and that it's a sixteen-mile drive from Gap Inc.'s actual corporate headquarters, in downtown San Francisco, as well as a fifteen-mile drive from Gap Inc.'s third San Francisco–area corporate "campus," a 283,000-square-foot building in Mission Bay. (Gap operates a shuttle-bus service between the locations. The buses reduce the number of intercampus solo car trips by employees, but they nevertheless represent a significant environmental cost—no bus is as green as an elevator—not to mention a waste of employees' time.)

In 1999, McDonough, speaking of the grass-covered roof on the Gap building in San Bruno, told *Time,* "Our idea was that if a bird flew over the building, it would not know that anything had changed,"[22] a brazen bit of media-directed greenwashing. Any bird with half a brain would see what you or I would see: a low-rise building surrounded by other low-rise buildings in a virtually unbroken expanse of parking lots, freeways, and suburban tract houses. Green roof or not, the Gap complex increases the embodied inefficiency of the carpet of sprawling low-rise development which now extends, virtually without interruption, from southern San Francisco to San Jose and beyond. McDonough's Gap building has plenty of individual features that are worthy of imitation by other designers, but its non-dense configuration and transit-free location make it just one more Automobile Objective.

Creating building roofs covered with vegetation can be a reasonable environmental strategy, because vegetated roofs can reduce heating and cooling needs, extend roof life, moderate noise transmission, and help to slow or eliminate the flow of rainwater runoff into overtaxed municipal storm sewers. (Vegetated roofs on buildings of all sizes have a potentially useful role to play in New York City, where storm-sewer loads are a significant problem.) But they are not important as repositories of plant species—a claim of McDonough's—and they can have a significant downside. They are expensive, they are not maintenance-free, and they require adequate structural support. (Rainwater weighs eight pounds a gallon, and one of the purposes of a green roof is to accumulate that weight and hold it in place.) Most important, a green roof should not be thought of as an environmental end in itself. Making a building shorter and wider in order to create more room for a rooftop lawn is LEED brain at its most demented. In Gap's case, the environment would have been better off if the company had decided to let native grasses fend for themselves and had, instead, consolidated its entire San Francisco–area workforce in a taller building with a smaller roof (vegetated or not) in a single, transit-accessible location downtown.

THE OPENING SENTENCE OF A RECENT ARTICLE IN *The Wall Street Journal* points to one of the most troubling consequences of the spread of LEED brain: "Most homeowners like the idea of going green—until they get the bill." LEED has helped to foster the widely held impression that reducing the environmental impact of human living spaces is largely a matter of buying more fancy stuff. "Being earth-conscious isn't always easy," the article continued. "Anna and David Porter decided three years ago to trade in their 4,000-square-foot Seattle home for a smaller, greener abode. They paid about $300,000 for an old house on a beachfront lot in Stanwood, Wash., and budgeted $450,000 to renovate it into a green showplace, with kitchen countertops made of recycled glass and concrete, a geothermal heat pump, a tankless hot-water heater, a solar electric system and cabinetry and flooring made from sustainably harvested wood."

One thing led to another, and the Porters ended up tearing down the existing beachfront house and then spending $1.2 million to build "a custom-designed, 2,700-square-foot replacement" on the same lot. Including the cost of the house they tore down, that works out to $555 a square foot—an exercise in conspicuous consumption and waste rather than an investment in any defensible conception of greenness. Yet the Porters were really just following the examples of the Rocky Mountain Institute, the Chesapeake Bay Foundation, William McDonough + Partners, and the like.

As Anna Porter herself told the *Journal*, the new house just seemed like "the right thing to do."[23] Yet this is exactly wrong. The best strategy for making a new single-family house greener is to build it on a small lot in an already dense neighborhood (which increases embodied efficiency), to build it smaller (which consumes fewer resources during construction, requires less energy forever, and discourages the accumulation of unnecessary possessions), to caulk and insulate it more thoroughly, especially under the roof (which helps to keep heat on the correct side of the building envelope in all seasons), and to go easy on the air-conditioning and the inefficient appliances. Taking those boring but highly effective steps dramatically reduces not only energy use and carbon output but also, in most cases, upfront construction costs and ongoing operating expense, even if it doesn't attract the attention of newspaper reporters or earn much in the way of LEED credits. And if you want to try for your own personal platinum level, you can make it an apartment in a mixed-use urban neighborhood served by public transit instead of a single-family house on the beach.

Even in dense urban areas, environment-related media attention tends to focus not on the structures with the highest levels of true embodied efficiency but on the ones with the largest collections of expensive eco-gadgets. In 2008, *The New York Times* featured a supposedly green townhouse in Chicago, a lavishly renovated former warehouse, whose owners had clearly spent a fortune adding what the article described as "an exhaustive list of green amenities" and what one of the owners described as "a complete set, a truly encyclopedic wonder cabinet of devices"— wind turbines, photovoltaic panels, geothermal wells, rainwater collection cisterns, dozens of others. The owners and their architect, as quoted in the article, sound almost awed by what they seem to view as their great contribution to the future of the civilization. "When people ask me why I have those wind turbines, I always wonder why they don't have them," one of the owners told the reporter. "It's like when Thoreau was in jail for an act of civil disobedience and Emerson visited him. 'Henry,' Emerson said, 'Why are you in jail?' To which Thoreau replied, 'Ralph, why are you not in jail?'" But what the house mainly represents is a highly developed ability to rationalize big-ticket purchases. There isn't enough wind on the site to justify the wind turbines or enough sun to justify the photovoltaic panels, and the lot is too small to allow for a truly effective geothermal system. Meanwhile, the architect turned the entire rear wall of the house into a modernist expanse of glass and galvanized steel, two dreadful insulators. "You've got to be doing it for other reasons," the architect said. But the environment doesn't gain from wasteful investments in inappropriate technology. Installing a pointless rooftop wind turbine is as wasteful as driving an SUV.[24]

LEED brain is really just another manifestation of our well-documented tendency, as the world's most ravenous spenders, to think of human fulfillment mainly as the sum of the things that we can buy. Another telling example is the house of Richard and Maryann Ellenbogen and their two children, in Pelham, New York, a suburb of New York City. The house was described in a 2008 Westchester newspaper article as "environmentally friendly" and as an "ultra-energy-efficient 'smart house.'" It does have the usual headline-grabbing big-ticket green gizmos—a geothermal-supported heating and cooling system, a large bank of photovoltaic panels mounted on a stone wall in the back-yard—but it also has a long list of extraordinarily wasteful and inefficient features, among them 8,000 square feet of interior space, five bedrooms, six bathrooms, a double-height-plus ceiling in the barn-size living room, a synthetic-sandstone fountain in a cavernous foyer, a fitness room, and a vast kitchen, which contains a "full arsenal of state-of-the-art equipment . . . from the Sub-Zero refrigerator and wine cooler to the Wolf stove, convection oven, and warming drawer to the two dishwashers." The house was funded partly by U.S. and New York state tax-payers, who paid $7,000 and $40,000, respectively, in the form of tax credits and rebates, toward the cost of some of the house's putative energy-saving amenities. Richard Ellenbogen, who is the president of a plastics manufacturing company, told a reporter, "It feels great to go home every night and know that the house is big and beautiful but that it's not detrimental to the environment."[25] This is, if nothing else, a truly unsustainable level of self-delusion.

When homeowners (and large corporations) decide to go green, the first major step they contemplate is often the addition of photovoltaic panels. Pulling free, emission-less electricity out of the sky is an irresistible idea, and solar panels have the additional appeal of making a conspicuous statement to neighbors and passersby, while also usually qualifying for significant tax breaks or other subsidies. But solar power is not as straightforward as people usually think. In Washington, D.C., in 2007, I attended a presentation by Steven J. Strong, who is the president of Solar Design Associates, a firm in Cambridge, Massachusetts. Strong is a thoughtful and knowledgeable clean-energy evangelist, with decades of direct experience in solar construction projects, and he made an impassioned case for solar power. But he also described solar as "the last thing I want to see you do." His point was that, in terms of environmental return on financial investment, the biggest gains almost always come not from installing photovoltaic panels but from such unglamorous but highly valuable steps as increasing the depth of attic insulation and using more efficient heating, ventilating, and air-conditioning systems. He described solar panels as "the dessert part," and described more mundane but more cost-effective energy-saving steps as "the vegetable part," and he urged his audience to tackle their vegetables first. The Pelham house, by this way of thinking, is almost all dessert. The environment (and taxpayers) would have been better off if the Ellenbogens had skipped the solar panels and shrunk their house to a reasonable size, lowered the ceilings, and cut back on the appliances. Taking those boring steps would have been far greener, although they wouldn't have gotten the house an article in the local paper.

A serious drawback of residential photovoltaic installations is that they tend to work best on buildings that, for other reasons, have very high levels of embodied inefficiency. Solar panels, to be truly effective, require unobstructed exposure to the sun, and that typically means they work best on widely separated buildings sited on big lots from which all intervening trees have been removed. This requirement, combined with the high price of even heavily subsidized panels, explains why residential photovoltaic installations tend to be most common on big, expensive houses far from urban cores. (It also explains why the installations are often described as "demonstrator" projects, as is the case with the Chicago couple and their remodeled warehouse.) Photovoltaic systems also raise extremely difficult issues regarding what happens when the sun isn't shining—as it usually isn't. The only truly efficient solar-storage medium which has been discovered on earth so far is fossil fuels—which are repositories of energy that originated on the sun and can therefore be thought of as easily portable solar batteries that don't lose their charge even after being buried underground for tens of millions of years. Accomplishing anything remotely similar with human technology has proven to be far more difficult. As of the end of 2007, solar energy from all sources accounted for much less than a tenth of a percent of total U.S. energy consumption. (Wind was a little better—about a third of a percent.)[26]

In 2007, I visited Natural Bridges National Monument, in southeastern Utah, a park remote from any existing electric grid. For most of the past thirty years, Natural Bridges has used a photovoltaic system to produce almost all the electricity that's used in its small compound of buildings, including its staff housing. It is literally a demonstrator project—it was selected by the U.S. government for that purpose in the late 1970s because of its isolation, its modest size, and its optimal solar exposure— but unlike most solar demonstrator projects it comes fairly close to being self-sufficient. To make the system work, though, the park had to:

- cut its pre–solar power consumption (which had been supplied entirely by diesel generators and had therefore already had to be cut to very modest levels in comparison with that of a typical suburb) by two-thirds;
- install a large and expensive storage system, which today consists of thirty-nine 1,200-pound lead-acid batteries housed in a climate-controlled building;
- install large backup diesel generators for times when the sun stays hidden for more than two consecutive days, the longest period the batteries can cover;
- power all heating and cooking with liquefied petroleum gas, which, along with the diesel fuel burned by the generators, must be delivered by truck. (Residents also drive their own gasoline-powered vehicles.)

Even at Natural Bridges, in other words, solar power is definitely dessert, because it supplies less than a third of the site's pre-1980 energy demand, and can't function on its own, without extensive backup.

Solar advocates often say that photovoltaic systems work better in locations where they can use the existing power grid for "infinite storage," obviating the need for batteries and backup diesel generators. The idea is appealingly simple: when the sun is shining brightly and the panels are producing more current than their owner needs, the extra power is fed into the grid, causing the owner's electric meter to run backward; at night and on cloudy days, when the panels aren't producing any current at all, power is drawn the opposite direction. This is known as "net metering," and it's permitted in a number of states. It enables a photovoltaic user to treat the grid as a sort of bank, depositing electrons when the sun is shining and withdrawing them when it's not. In ideal circumstances, the deposits and withdrawals cancel each other out, leaving the owner with an electric bill of zero and the utility with an additional source of current.

Naturally, it isn't this simple in reality. The main difficulty is that the grid can't truly act like a bank, because the excess power provided by solar panels isn't necessarily fed into it at times when excess power can be used. The reason has to do with how most electricity is generated and consumed. Every electric utility has what's known as a base load—the minimum level of power demand that goes on all the time, twenty-four hours a day, all year long. The utility generates electricity to meet this load with big plants, usually powered by coal or natural gas, that produce electricity at very close to the same rate all the time and, for the most part, are never turned off, since the plants can take a long time to shut down, restart, and bring back to capacity. An electric utility also has what's known as a "peaking load," which is power demand above and beyond the base load. Peaking load is quite variable; in most parts of the country, it tends to reach its maximum each day late in the afternoon, around four or five o'clock, when people are getting home from work and school and making themselves busy around their houses, yet other people are still finishing up at the office, making copies and turning on lights— and it's at its highest on the hottest days of the summer, when returning homeowners are the most likely to crank up their air conditioners. Utilities meet these variable loads with smaller, supplemental generating facilities, which are far less efficient than base-load plants but can be fired up and shut down on much shorter notice. In order to do this effectively, the utilities closely monitor not only electric usage but also weather forecasts and other external factors, so that they can smoothly add and subtract capacity in close anticipation of likely demand. Any power-generating system must always be in balance in this sense, because producing current in excess of demand is a waste, while failing to meet demand leads to voltage drops and outages.

The difficulty with solar net metering is that the additional current that is provided to the grid by residential photovoltaic panels often arrives at times when the grid is unable to use it: the solar peak generally occurs around midday, often several hours before the electric-demand peak, and the extra electrons often have nowhere useful to go. If the sun shone fully at the same time every day for a predictable number of hours, utilities could scale back their base generating capacity and let widely dispersed residential solar installations take up the slack, but the weather doesn't work that way. Since most residential customers would be unwilling to endure power failures and brownouts at times when the sun didn't happen to be shining at all, utilities can't necessarily cut back their base generating capacity. As a result of these and other factors, net-metered solar installations sometimes serve mainly to increase the electric bills of other customers—since the utilities end up paying for current they can't use, in addition to subsidizing the installation of the panels in the first place. In most parts of the country, wind turbines pose a similar dilemma. Wind is even less predictable than sunshine—which means that sudden changes in intensity are far harder to compensate for with peaking generation, since utilities usually can't anticipate them precisely—and the days of the year when the extra power generated by turbines would be the most useful to the grid tend to be hot summer days when, almost by definition, power-generating breezes are not blowing. For these and other reasons, isolated photovoltaic and wind installations almost always provide less useful wattage than their theoretical generating capacity would suggest. They also have operating costs that are obscured by the subsidies which support them, and that are often overlooked in discussions of their environmental utility. Every system, for example, requires an inverter, an expensive piece of equipment that turns the direct current produced by the system into the alternating current required by electric appliances, and which has to be replaced every five or ten years. Photovoltaic systems also gradually decline in efficiency and have a definite useful life span, usually twenty or twenty-five years. In January 2008, Severin Borenstein, who is a professor at the Haas Business School, at Berkeley, and the director of the University of California Energy Institute, concluded that, based on "moderate assumptions" about interest rates and the cost of electricity, "the net present cost of a solar PV installation built today is three to four times greater than the net present benefits of the electricity it will produce."[27] This doesn't mean that the sun won't be a critical energy source going forward. But it does suggest that we can't solve our energy and emissions problems by putting a few solar panels on everyone's roof and proceeding as before. (The biggest source of energy at Natural Bridges, by comparison with the period before 1980, is not solar panels but forced conservation, in the form of the park's two-thirds reduction of its pre–solar power demand—a form of virtual energy that Amory Lovins has usefully named the "negawatt.")

A further complicating factor regarding all forms of electric power is that demand for electricity in the United States is certain to change radically in coming years. There has been much talk, for example, of replacing more and more gasoline-powered cars with electric cars (which run on rechargeable batteries) and with so-called plug-in hybrids (which run on rechargeable batteries when possible and switch to a gasoline engine when the batteries are depleted). This seems likely to happen, on some scale. But there are many unresolved issues, including the fact that physics, so far, has stood in the way of creating batteries that can safely power a car for more than thirty or forty miles without being recharged. Even more important is the question of where the electricity required to recharge all those high-capacity car batteries will come from. Popular discussions of electric cars usually end at the wall outlet—just plug 'em in when you get home from work, and off you go again in the morning!—with no consideration of how much new generating capacity will be required to meet this potentially huge new demand, or when in the daily generating cycle it is most likely to occur. (One driver plugging in his car when he gets home from work is not a problem; a million drivers plugging in their cars when they get home from work—at or near the moment of peak daily energy demand—would strain the power grid; an entire nation of electric-car drivers would require a total American energy makeover.) Electric cars and plug-in hybrids will increase, not end, our reliance on fossil fuels, because they will add to the nation's overall electric load and thereby increase our need for generating capacity, which in the United States, for decades to come, will be powered mainly by burning coal or natural gas.* Swapping one energy source for another does nothing to solve our underlying energy dilemma. As the experience of Natural Bridges suggests, the power we don't use is more important than the power we do. We must significantly reduce the number of miles we drive, not merely replace one motor fuel with another one.

THE ELLENBOGEN "GREEN" HOUSE IN PELHAM HAS large double-paned windows filled with argon, an inert gas— high-cost, high-performance windows of a type that has been promoted so heavily in recent years that many consumers have

---

*Some American drivers have even more ambitious plans for natural gas. Internal combustion engines can be made to run on compressed methane, and this fact is currently being promoted as an our-prayers-are-answered solution to the American driving crisis. A recent two-page newspaper advertisement by Chesapeake Energy Corporation describes compressed natural gas as "clean, abundant, affordable, American," and says, "More than 70% of U.S. homes have access to natural gas. With a simple compressor unit installed in the garage, drivers can refuel overnight." Best of all, the advertisement says, compressed natural gas costs 40 percent less than gasoline and pollutes only 10 percent as much. All of this is misleading. Natural gas prices fluctuate seasonally, sometimes severely. The United States has been a net importer of natural gas for decades, mostly from Canada, and our import needs will grow as demand for gas rises. (Meanwhile, of course, Canada's ability to supply us will fall as demand there rises, too, especially if Canadian drivers also develop a taste for methane-powered cars.) We already consume roughly 23 trillion cubic feet of natural gas a year; that's the energy equivalent of something more than half of current U.S. oil consumption. Converting any significant portion of the American motor fleet to natural gas would quickly send prices higher. The short answer is that, with energy, there's never a free lunch. We're going to burn lots and lots of natural gas in coming years, but it's not the secret permanent answer to our troubles.

gotten the idea that the more gigantic windows a house has the more energy-efficient it must be. Indeed, the supposedly sustainable houses and office buildings depicted in news reports often seem to be made mainly of glass. In 2007, *Wired* magazine conducted an extensive promotional campaign for what it described as an eco-friendly "house of the future," which had been built on a winding residential street in Brentwood, California, a suburb of Los Angeles. The house, which was designed by Ray Kappe and prefabricated by a green-oriented factory builder called Living-Homes, has five bedrooms and four and a half bathrooms, and it has various putatively green features, among them soy-based foam insulation, a rooftop photovoltaic system, bathroom countertops made of recycled glass, and ceilings paneled in reclaimed wood. It also, typically, has exterior walls that are mostly window.[28]

Windows are a lot more energy-efficient than they used to be, but even very expensive, high-tech windows are poor insulators, in comparison with other building components. Sunlight streaming through large glass surfaces fights air conditioners during hot weather, and heat escaping through large glass surfaces undercuts heating systems during cold. Some of the huge windows in the *Wired* house are equipped with elegant-looking gauzy blinds, which can be pulled down on hot, sunny days, but all that glass still represents a significant source of heat gain during the day and of heat loss during the night. Nevertheless, as the *Wired* house shows, many people have been led to believe that a good way to make a house greener, or appear greener, is to use more glass. Big glass walls have a clean, modernist look, consistent with popular impressions of environmental responsibility, but using more glass necessarily means using less insulation, and, as the Department of Energy has explained, "structures with high glazing areas are less likely to comply with the energy code."[29]

Another reason for the belief that glass is green is that the most widely discussed green structures in recent years have tended to be office buildings, which often have lots of glass for reasons that mainly have to do with aesthetics. Office buildings also have energy-use patterns that are different from those of most residences. As much as half of the electricity used in a typical commercial building goes for lighting, and, because such buildings are used mainly during daylight hours, increasing the amount of ambient sunlight in interior work spaces can significantly reduce total electric loads. Still, such buildings usually have far more window surface than the trade-off between lighting efficiency and thermal efficiency calls for, and designers have to compensate for the resulting heat gain and heat loss by beefing up cooling and heating systems and making other modifications. (This is most significant in hot climates, because computers and other office machines generate significant amounts of waste heat, adding to warm-weather air-conditioning needs.) In residences, lighting patterns are different, because houses are often empty or only sporadically occupied during daylight hours, and their highest levels of energy use typically occur late in the day and in the evening, once the occupants have come home from work and school and have begun making dinner and turning on lights, flat-screen televisions, computers, air conditioners, clothes dryers, and other power-consuming devices.

A third reason that big, expensive windows have an exalted environmental reputation is that window manufacturers have worked hard to give them one. The insulating value of many building materials—as you know if you've ever shopped for insulation—is indicated by a numerical rating known as an R-value, which is an expression of the material's ability to resist the flow of heat. The formula used to calculate R-values has to do with the time it takes a given amount of heat on the warm side of a material to raise the temperature of the cool side by a given amount; the calculations are complicated, but the basic point is that higher numbers are better than lower numbers. (An inch-thick sheet of expanded polystyrene—the stuff that packing peanuts are made of—is about R-4; a foot-thick layer of fiberglass insulation is about R-38 and is therefore more than nine times as effective as the thin polystyrene sheet at resisting the flow of heat.) All the elements of an enclosed structure—including not only insulation but also lumber, wall hangings, and the film of air adhering to interior and exterior surfaces—resist heat flow to some extent and therefore have R-values of their own, and you can determine the total thermal resistance of an entire system by adding up the numbers. A typical modern house wall—with Sheetrock on the inside, wood siding and sheathing on the outside, and five and a half inches of fiberglass insulation in between—has a total R-value of somewhere around

R-21, although R-values, like most single measures of complex systems, work better in broad comparisons than they do in precise evaluations. The true heat retention provided by any wall or other structural element, no matter what its stated R-value, is affected heavily by how conscientious the builders were about filling gaps and plugging air leaks, and by such external factors as how hard the wind is blowing and whether or not it is raining.

In windows, thermal performance is expressed not in R-values but in U-factors, which measure heat conductance rather than heat resistance, and with U-factors lower numbers are better. Builders, window manufacturers, and environmentalists almost always claim that U-factors are nothing like R-values. For example, the National Fenestration Rating Council, a nonprofit organization that tests and rates windows, states, in a publication titled "Fenestration Heat Loss Facts":

> Windows are different than insulation in walls and ceilings. Windows let the light in and allow people to see out, and they interact with their environment in ways that insulation does not. They react to outside temperatures, sunlight, wind, indoor air temperatures, and occupant use. Windows are affected by solar radiation and the airflow around them. R-value does not accurately reflect this interaction. Therefore, the window industry measures their product's energy efficiency with a **thermal transmission rating, or U-factor.** U-factor measures the rate of heat transfer through a product from hot to cold. Therefore, *the lower the U-factor, the*

lower the amount of heat loss, and the better a product insulates a building. . . . The biggest difference between U-factor and R-value is that U-factor measures the entire fenestration product's heat transfer rate while R-value measures heat loss resistance of a homogeneous material (a material comprised of one component such as insulation only). [All emphases are in the original.][30]

But all of this is misleading, and much of it is simply wrong. A U-factor doesn't contain any information that an R-value doesn't: each number is simply the reciprocal of the other. If you know a material's R-value, you obtain its U-factor by dividing its R-value into 1, and if you know its U-factor you obtain its R-value by dividing its U-factor into 1. Presumably, window manufacturers use U-factors not for the reasons enumerated by the NFRC but because they worry that the R-values of even the most energy-efficient high-tech windows would seem pretty measly to people trying to decide whether such windows are worth the considerable extra cost. An ordinary wood-framed single-glazed window with an aluminum storm window has an R-value of about 2, while a fancy wood-framed triple-glazed window with quarter-inch spaces between the panes (but no storm window) has an R-value of about 2.5. A triple-glazed window that has a low-emissive coating on the glass and is filled with argon or krypton instead of ordinary air, has an R-value of 4 or 5. That's a large relative improvement over an ordinary window, but it's a small absolute one; you could match the impact of one high-tech window on the overall R-value of a wall by adding an extra three-quarters of an inch of fiberglass insulation to one window-size section, and you could greatly exceed it by simply omitting the window entirely. And the saving in energy consumption from an entire room's worth of high-tech windows can be negated if you open one of them a crack in cold weather to admit fresh air, or fail to seal all the gaps around its frame during installation, or position it in a wall that doesn't receive direct sunlight during heating season (thereby reducing its potential contribution to daytime heating of the house) or that receives too much direct sunlight during cooling season (thereby increasing the need for air-conditioning). Also, argon and krypton tend to leak out of sealed windows (the molecules are very small). Window manufacturers sometimes say that such leakage—which cannot be detected—has little impact on a window's overall thermal performance, an assurance that might make a smart shopper hesitate to pay for it in the first place. Yet manufacturers and environmental groups have pushed such windows so heavily that many people view them as almost magical. An executive of the Natural Resources Defense Council had me place my hand on one of the fancy windows in his office, at that organization's offices in New York, and asked me if I wasn't amazed at how warm it felt. But that window, in comparison with the wall in which it was set, actually represented a relative energy leak.

The real-world value of expensive window add-ons can be very hard to judge. Low-emissive glass, for example, has a microscopically thin metallic coating that is intended to reflect some infrared and ultraviolet radiation while admitting as much visible light as possible. But the actual role it plays in energy efficiency depends on many factors, among them total glazing area, orientation of the windows in relation to the sun, geographical location of the structure, use of curtains, blinds, and awnings, presence or absence of shading vegetation, amount of dirt adhering to the coated surface, and whether the principal issue is heat gain or heat loss. For example, a coating that reduces solar heat gain during the summer, and therefore decreases the load on air conditioners, also reduces solar heat gain during the winter, and therefore increases the load on the furnace. In a house that doesn't have air-conditioning, low-e windows can actually increase energy use, by reducing the sun's effectiveness, during the winter, at passively heating rooms with southern exposure. And, as is often the case, windows and other building components often function differently in actual practice than they do in their manufacturers' claims and descriptions.

The focus on high-priced, high-tech windows reinforces the common perception that reducing residential energy consumption is a luxury upgrade, beyond the financial reach of the average person. But there are dozens of unexciting yet relatively inexpensive steps that even wealthy homeowners should take first, before investing tens of thousands of dollars in argon-filled glazing. No one, for example, should shop for fancy windows before beefing up attic insulation, cleaning or replacing the furnace, insulating the water heater, sealing gaps and air leaks around existing windows and doors, making better use of existing window coverings, and relearning the temperature-control tricks that people used to use instinctively back when there were no other options. Ann and I don't have central air-conditioning, but we have learned to keep the main parts of our house comfortable on all but the hottest summer days by doing what our parents and grandparents did: opening windows at night, to cool the entire house, then shutting the windows in the morning and drawing the curtains in the sunny rooms, to keep them from rapidly heating up again. Modern HVAC systems have made most of us lazy about temperature control, and therefore about energy use: when we feel uncomfortable, we adjust the thermostat rather than identifying the source of the problem and looking for a low-tech remedy. Installing high-tech windows, like installing rooftop photovoltaic panels, should be considered "the dessert part," in the sense that Steven J. Strong meant, and should be contemplated only once all the simpler and far more cost-effective steps have been taken. Thomas L. Friedman, in his recent book *Hot, Flat, and Crowded,* conveys this same basic idea in a memorable chapter title: "If It Isn't Boring, It Isn't Green"—seven words that should be adopted as a mantra by all environmentalists, as a reminder of the dangers and temptations of LEED brain.[31]

ON THE AFTERNOON OF AUGUST 14, 2003, I WAS WORK-ing in my office, on the third floor of my house, when the lights blinked and my computer's backup battery kicked in briefly.

This was the beginning of the great blackout of 2003, which halted electric service in parts of eight northeastern and midwestern states and in southeastern Canada. (My house and the rest of my town were unaffected after the initial flicker.) The immediate cause was eventually traced to Ohio, but public attention often focused on New York City: it had the largest concentration of affected power customers, and the lights there remained out for longer than they did in most of the rest of the blackout region. In the minds of many people, including many New Yorkers, one key to preventing future large-scale outages will be finding new ways to reduce the huge amount of electricity that New York City consumes.

This notion seems unassailable on its surface, since discouraging consumption is always good, of course. But it's wrongheaded, for the reasons I explained in the first chapter: people who live in cities use only about half as much electricity as people who don't. Richard B. Miller, who resigned as the senior energy adviser for the city of New York six weeks before the blackout, reportedly over deep disagreements with the city's energy policy, told me in 2004, "When I was with the city, I attended a conference on global warming where somebody said, 'We really need to raise energy and electricity prices in New York City, so that people will consume less.' And my response at that conference was, 'You know, if you're talking about raising energy prices in New York City only, then you're talking about something that's really bad for the environment. If you make energy prices so expensive in the city that a business relocates from Manhattan to New Jersey, what you're really talking about, in the simplest terms, is a business that's moving from a subway stop to a parking lot. And which of those do you think is worse for the environment?'" Attempting to address the nation's energy problems by targeting the people who, individually, draw the least amount of power is counterproductive. If every American used as little electricity as the average New Yorker does, total U.S. energy consumption would drop to a fraction of what it is today. A truly enlightened energy policy, therefore, would reward New Yorkers and other urban dwellers for living in places that force them to be parsimonious, and it would encourage other people to follow their good example.

Yet for the most part we do the opposite. New York City residents pay more per kilowatt-hour than do almost any other American electricity customers. The reasons for the difference are complex, but they include the fact that electricity is more heavily taxed in New York City than in almost all other parts of the country. For the city's residential and commercial customers, the cost of power is inflated by a long list of direct and indirect government charges, most of which are not enumerated on energy bills: sales taxes, excise taxes, property taxes, and gross-receipts and franchise taxes, among others. These charges—which provide significant revenues for the city and the state—can constitute close to 20 percent of the cost of power for residential and commercial users. Making electricity more expensive is an effective way to encourage conservation, but concentrating the surcharges in areas where per-capita use is already unusually low makes no sense, economically or environmentally.

It's not just taxes that penalize efficient urban dwellers. In sprawling areas, the cost of providing electric service to far-flung new subdivisions is typically spread across the entire utility customer base, giving no one an incentive to make development more compact and efficient. Most civic infrastructure in the United States is treated in this irrational way. Developers build on cheap land far from cities' central cores, and then power lines and roads and sewers and water mains and new schools follow to serve them, raising costs for efficient and inefficient users alike. No one who moves into a new suburban subdivision pays anything like the real cost of the infrastructure that is required to support them. Low-volume energy users in central cities end up subsidizing the expressways, air conditioners, swimming-pool heaters, minivans, and automatic lawn watering systems that serve people living at the farthest edges of the region; a rational system would do the reverse, by imposing penalties for squandering resources shared by all.

Richard Miller, after leaving his job with the city of New York, went to work for Consolidated Edison, the principal electric utility in the New York metropolitan area, a move that surprised a number of his friends but was driven by his thinking about the environment, and by his belief that state and local officials have historically taken unfair advantage of the fact that there is no political cost to picking on large utilities. Con Ed pays about a billion dollars a year in property taxes, making it by far the city's largest property-tax payer, and those charges inflate electric bills. Meanwhile, the cost of driving is kept artificially low. (Fifth Avenue and the West Side Highway don't pay property taxes, for example, and drivers aren't charged for using them.) "In addition," Miller told me, "the burden of improving the city's air has fallen far more heavily on power plants, which contribute only a small percentage of New York City's air pollution, than it has on cars—even though motor vehicles are a much bigger source."

Miller's career path has been heavily influenced by his growing conviction that most people, including most environmentalists, see many important energy issues upside down. After graduating from Amherst College, in 1980, he spent two years as a Peace Corps volunteer in West Africa. "When you live without electricity for two years, you see how much electricity means to the quality of people's lives," he told me recently. "You also see a lot of the inefficiencies and environmental degradation that result from living without it. People tend to think that living a simple life in Africa can't possibly be bad for the environment, but if you live in a borderline desert area and don't have electricity or natural gas, then you have to chop down trees, because you have to burn wood to cook." After the Peace Corps, Miller went to law school, specialized in energy law, and eventually moved to Brooklyn. He and his wife and their two children have lived without a car for ten years.

"The primary drivers behind growth in electricity use are population growth and economic growth," he continued. "When those things occur in a central city, like New York, instead of in the suburbs or the exurbs, it's good for the environment, not bad. And if we need to invest in more power plants or increased distribution capacity to meet the needs of a central city, then we should do it, because when growth takes place there we know that energy is being used more efficiently than it could be otherwise."

This is by no means a widely shared belief among environmentalists, who have generally looked upon big utilities with no more affection than they have for big cities. A growing theme in American environmentalism, in fact, has been the encouragement of "distributed generation"—the decentralizing of the production of electricity by supplementing, and in some cases circumventing, the existing power grid through the creation of smaller and more widely dispersed power-generating facilities, including ones that serve individual buildings. A prominent proponent has been Amory Lovins of RMI, who has tantalizingly described a future in which automobiles are replaced by hydrogen-powered "hypercars," and homeowners generate their own electricity with photovoltaic arrays on their roofs and air-conditioner-size hydrogen-powered fuel cells—which generate electricity without combustion—and produce their own hydrogen, with in-home "hydrogen appliances" fed by natural gas. (Hydrogen fuel cells produce electric current by running electrolysis in reverse: instead of using electricity to split water molecules into hydrogen and oxygen—a standard high school science experiment—they combine hydrogen and oxygen, in the presence of various catalysts, to produce water and electricity.)

These ideas have extraordinary emotional appeal, and Lovins has promoted them all over the world. (He's a coauthor of a thick book on the subject, *Small Is Profitable: The Hidden Economic Benefits of Making Electrical Resources the Right Size.*[32]) But the full Lovins concept depends on, among other things, finding solutions to the staggering problems associated with using hydrogen as a fuel—not the least of which is devising a method of extracting it which doesn't consume more energy than can later be provided by the hydrogen itself. Nor are fuel cells the panacea they are often claimed to be. (In 2004, a man who called in to a public radio show on which I was being interviewed said that cars in the near future would not require "energy" in any form, because they would run on water—a profound misunderstanding of fuel-cell technology.) Even when fuel cells have been used—almost always in demonstrator applications, since their extremely high cost makes them uneconomical in comparision with conventional sources of power—there are complications. One of the environmental goals of the designers of 4 Times Square was to "minimize the amount of transmission loss and to offset the huge electrical load of the Times Square signage* by generating electricity with fuel cells," as the architect Daniel Kaplan explained in 1997. But

---

*One of the many peculiarities of New York City zoning rules is that buildings in the Times Square district are *required*, for historical reasons, to have massive, wasteful illuminated signs. (Broadway has been called the Great White Way since the early 1900s because of its dazzlingly bright theater marquees and billboards.)

doing so was far from straightforward. "Each fuel cell [of a total of eight] requires approximately 700 square feet of floor space," he continued. "No open land was available as the building is built up to the property line. It was not advisable to locate within the building as the units give off heat and CO_2. The units need major overhaul every five years, requiring replacement of a 9,000-pound part. Once the units are turned on, it is inefficient to turn them off, but the building cannot use all of the power generated during very late night hours. Economic use of the excess power is an issue."

The designers' solution was to put the fuel cells on the building's roof and add a built-in crane system for hauling replacement parts up from the street, seven hundred feet below. As for the excess power, the solution was to sell it to Con Ed by feeding it into the grid—although that solution merely shifted the economic problem into someone else's pocket, since the grid doesn't need excess power late at night, either, because that's when the city's loads are the lowest. If fuel cells make economic sense, it would be more rational and efficient, in dense cities, to incorporate them into the local utility's base generation capacity and run them in the opposite direction, feeding power from the grid to electricity customers. But it's not clear, yet, that they make sense even there.

Some of the main difficulties behind the goal of decentralizing power production were enumerated in 2000 in an article in *Electricity Journal* by Nathanael Greene and Roel Hammerschlag, of the Natural Resources Defense Council. "Given the size of the electricity market, the range in emissions and the lack of regulation in this area," they wrote, "there is clearly the potential for a literal thousand points of light to become a thousand points of soot. If just one half of one percent of the U.S. demand for electricity were met by uncontrolled diesel engines, the country's annual nitrogen oxide emissions could increase by nearly five percent."[33] Lovins and other proponents of decentralizing power production aren't promoting diesel generators, of course, but most of the small-scale, non-grid energy production in the United States actually is diesel-powered. Large generating plants are inherently more efficient than small generators; they also do less damage to the environment per unit of output, since "fitting plants with best available control technology can be financially feasible on a large scale, but not on a small scale." Greene and Hammerschlag are by no means dismissive of distributed generation, but their paper repeatedly demonstrates that no idea can be judged apart from its real-world context. Heating New York City apartments with individual woodstoves rather than with steam from Con Ed's centralized steam distribution system—which supplies most buildings in Manhattan below Ninety-sixth Street, and is used to co-generate electricity—would not be a gain for the environment.

The existing American power grid is antiquated, and it is vulnerable to failures like the one that shut down power in the Northeast in 2003, but breaking it apart is unlikely to be the solution, especially if doing so encourages people to live even farther from one another, thereby increasing wastefulness and environmental damage of all kinds. Part of the fascination with distributed generation, and with the dream of hydrogen as an energy panacea, arises from our very worst impulses—the same yearnings for personal independence that lead to expressways, strip malls, and sprawl. (What is a car but distributed transportation?) The desire to produce your own power in your own basement is akin to the desire to drive yourself to work and swim in your own pool and play tennis on your own court: to be liberated from the grid is to be liberated from other people. That's a potent element of the American psyche, but it isn't remotely green, because it penalizes population density and encourages forms of development that are inherently destructive and inefficient. The surest, greenest way to shrink a power-generation-and-distribution system is to move its customers closer together.

Six

The Shape of Things to Come

Until relatively recently, we Americans have had our extraordinarily wasteful lifestyle mainly to ourselves, but large parts of the rest of the world are intent on catching up, with profound consequences—for us, for them, and for everyone else. Many Americans view China and India with personal alarm, since those countries' residents are making increasing use of dwindling natural resources that we've grown accustomed to thinking of as ours. Their economic growth and increasing reliance on automobiles have quickened the pace at which global oil supplies are diminishing, and, as a consequence, have raised our cost of living. They have also increased the rate and scale of global environmental damage. Many American consumers, in fact, think of the Chinese as the leading villains in our present energy, environmental, and economic crises, because there are 1.3 billion of them and their oil consumption more than doubled during the decade between 1996 and 2006. (By contrast, U.S. oil consumption during the same period grew by less than 10 percent.)

But China's growth figures obscure a more important point. The average Chinese still consumes only about a quart of oil per day, or 9 percent as much as the average North American, and that means that in 2005 the Chinese were just the 128th-largest per-capita oil consumers in the world.[1] Their rank is rising, but the Chinese, individually, have a long way to go before overtaking Americans and Canadians. Chinese per-capita consumption of just about everything is still only a fraction of ours, and a significant portion of China's energy use and production of greenhouse gases arises from manufacturing goods to sell to us less expensively than we could manufacture them for ourselves. It's hardly reasonable, therefore, for us to view the Chinese, or any other non–North Americans, as the cause of our present difficulties—just as it would be imprudent for the Chinese to be unconcerned about American economic problems, since China's prosperity depends in large measure on our continued ability to consume what its factories produce.

Nevertheless, if America's growing energy-and-emissions predicament proves anything, it's that an automobile-dependent society is vastly easier to create than to un-create. Moving from walking, bicycling, and transit to driving is relatively simple, because it requires only wealth, a desire for independence and status, and an inability or an unwillingness to look very far into the future; moving from driving back to transit, bicycling, and walking is far harder, because the cars themselves are only part of the problem. Much more critical is the embodied inefficiency of the way of life that cars make both possible and necessary, and of the sprawling web of wasteful infrastructure that high levels of individual mechanized mobility lead affluent societies to create. China, India, Brazil, and other countries whose economies have grown rapidly in recent years are following a path whose course and destination we are uniquely positioned to predict. They'd be better off, in the long run, if they took advantage of our experience and, by studying the evolution of our current dilemma, anticipated the difficulties that they, too, must eventually face. Reconfiguring society to suit the automobile was an unfortunate but probably unavoidable transformation in the United States in 1920; it's a worse but less inevitable one in China and India in 2010. There is a widespread feeling, even in the West, that it would be unfair to suggest that the citizens of less prosperous countries should forgo conveniences and luxuries that we Americans have come to think of as our birthright. But it's really the Chinese and the Indians who have the most to lose by following the worst elements of our century-old example. Sooner or later, no matter what else happens, the world will run out of affordable oil. Countries with expanding economies would be better off if they used their new wealth to create ways of life that they might hope to sustain beyond that inescapable end point, rather than recklessly investing in a future without a future. Not jumping off a cliff is easier than turning back in mid-fall.

Unfortunately, though, China, India, and many other countries seem intent on betraying their own long-term self-interest. India is a decade into one of the largest road-building projects ever undertaken, a 3,600-mile superhighway known as the Golden Quadrilateral, which links the country's four largest cities, plus an extensive network of feeder roads. Those new highways, in combination with India's brand-new "People's Car," the $2,500 Tata Nano, represent an environmental, economic, and cultural disaster in the making, because they are helping to propel the world's second most populous country toward a collision with the end of cheap oil. That is bad news not only for India but also for us and the rest of the world, since all our fates, increasingly, are intertwined. The last thing the world needs is to extend its dependence on automobiles. The fact that the Tata is small and fuel-efficient only makes its environmental impact worse, since those qualities will ease its penetration into places where cars have always been rare and might never have reached otherwise—places that now, inevitably, will evolve to suit the needs of drivers. The Tata is a global virus on wheels.

China's headlong rush into prosperity is at least as alarming. China is undergoing what one writer has called "the fastest motorization in history,"[2] and during the past dozen years it has achieved what another writer has called "the largest and most sustained economic expansion"[3]—in both cases, with environmental consequences still to be determined. There are currently about 40 motor vehicles for every 1,000 people in China. That rate of ownership is just a twentieth as high as that of the United States, but until the recent collapse of the global economy the Chinese motor fleet was expanding at a vastly faster pace.[4] We owe much of our own recent prosperity to the Chinese, who have supplied us with hundreds of billions of dollars' worth of inexpensive manufactured goods while simultaneously helping to finance the debt we incurred to buy them. Unfortunately, the Chinese also seem determined to repeat and perhaps top some of our most regrettable mistakes.

I VISITED CHINA IN NOVEMBER 2006. TRAVELING BY taxi from Beijing's airport to my hotel in the rapidly expanding center of the city, I felt like Esther Summerson, in *Bleak House*, approaching London for the first time: "I was quite persuaded that we were there, when we were ten miles off; and when we really were there, that we should never get there."[5] The air was pretty Dickensian, too. During my visit, the smog became so thick that flights were delayed, sections of six expressways were closed, outdoor school activities were canceled, three people were killed in traffic accidents in which reduced visibility was believed to have been a contributing factor, and buildings a block distant from my hotel were discernible only as dim silhouettes in the brown miasma. Air quality in the United States has improved so much in the past three decades that most

Americans no longer think of old-fashioned air pollution as an environmental issue; Beijing provides a useful reminder that carbon dioxide isn't the only industrial emission the world needs to worry about. Mustard-colored wisps even seemed to infiltrate my hotel room, although I may just have been imagining that. Nevertheless, I eventually came down with a brutal cold, which hung on for a month after I got home. "This is unusual for November," a Beijing resident told me on one of the smoggiest days. "Ordinarily, the air only gets like this in August."

August was on residents' minds because the Olympics were coming in August 2008, just a year and a half in the future. Preparations were under way all over the city. Men were being urged to stop spitting in public. Young women wearing dark uniforms stood near escalators in many public buildings and said, "Please watch your step," in careful English as passengers approached the bottom—practicing for similar assignments during the games. Most conspicuous of all, new buildings were going up everywhere. Immense construction cranes loomed in the smog. Someone told me that half of all the world's big cranes were in China at that moment, and his statement seemed as likely to be an underestimate as an exaggeration.

The rapid modernization of parts of China has created tremendous tensions between the past and the future. Some of the new buildings I saw were among the most structurally advanced high-rises in the world, yet on those same construction sites I often heard a sound that is seldom heard anymore on similar sites in the United States: the sound of many people hammering.

Labor is cheap and plentiful in China, and most work sites are notable for their large numbers of minimally skilled laborers doing things like carrying small loads slung from poles, using sledgehammers and foot-long cold chisels to break up macadam, dangling from bamboo scaffolds, riveting girders while wearing ordinary street shoes or even soft-soled slip-ons, and transporting bricks a hundred yards by dragging them six at a time on the blade of a shovel. A few days after I arrived in the city, I returned to the airport to take a tour of its huge new terminal, which was then nearing completion. My guide was Rory McGowan, a structural engineer with the British engineering firm Arup, which had worked on the building and has been deeply involved in many of China's biggest construction projects. McGowan was born in Ireland in 1964, and since 1986 he had worked for Arup all over the world. He had lived in Beijing for a year and a half with his wife, who is Russian, and their two sons, who were five and eight at the time of my visit and were already able to holler in Mandarin at Chinese tourists who tried to photograph them as ethnic curiosities. Inside the terminal, McGowan showed me some very ragged-looking concrete steps and said that in China it made economic sense to pour concrete quickly and sloppily and then pay inexpensive workers to chip off the lumps and fill in the voids. Outside the terminal I saw many acres of new concrete taxiways, and McGowan said that they had been poured one small section at a time and finished by hand, rather than being rolled out like carpeting by enormous concrete-placing machines of the type that unspool new taxiways at big American airports. That day, workers were pecking at the freshly hardened surface with hand tools, and sweeping up tiny piles of concrete dust with brooms made of twigs. (China is the world leader in futile sweeping.) Around the edges of the site I saw many other people, not officially working on the project, who were scavenging broken bricks, scraps of wood, and small pieces of twisted metal, and loading them onto already overloaded bicycles.

These were glimpses, still plentiful, of the China that is passing away. Once, not long ago, Beijing was almost entirely a city of bicycles, a form of transportation to which it is ideally suited, since the terrain is mostly flat and easy on the legs. Now, rapidly, it is becoming a city of cars, and many of its streets are actually closed to cyclists, and function as uncrossable barriers to pedestrians. Cars are so new in China that drivers haven't had time to evolve rational driving behavior, a comprehensible system of automotive etiquette, or even a fully developed theory of right-of-way. On some of the big roads in downtown Beijing, the exit ramps are not fully separated from the entrance ramps, and the result is that cars trying to get off have to compete with cars trying to get on. (Something similar happens on Chinese trains and elevators, where the idea of allowing the disembarking passengers to exit before the boarding passengers squeeze on has yet to take hold.) At busy intersections, Beijing drivers seldom wait for traffic to clear before attempting to turn left: they inch forward into the opposing stream of cars until they have intimidated a driver into stopping, then take control of the intersection until the halted driver on the other side has inched forward far enough to take the intersection back. During rush hour one evening, my taxi—which, characteristically, had no seat belts and had a manual transmission that made an excruciating grinding noise every tenth or twelfth gear change—was waiting in the left-turn lane on a busy six-lane street near my hotel, and when the traffic light turned red the four cars directly ahead of us, which were also waiting to turn left, immediately cut across the three opposing traffic lanes and got *inside* the cars now turning right from the perpendicular cross street to our left—and from there they set about bluffing their way to the other side. Of course, this tactic made traffic in both directions move more slowly than it would have if everyone had waited his turn, but the local drivers didn't see it that way—partly, perhaps, because driving was so new to them, and partly because line-cutting has deep roots in Chinese urban culture. On the day I arrived in Beijing, a young Chinese woman just behind me in the taxi line at the airport gradually moved up beside me and then, when the line made a switchback turn around a stanchion, passed me on the inside. She wasn't embarrassed to have done this, and she apparently didn't feel my eyes burning holes in her back for the next ten minutes.

One day, with McGowan and his family, I visited a "wild" section of the Great Wall, well away from the reconstructed portions that tourists usually see. Our car trip back to Beijing that evening took us down a curving, two-lane mountain road, on which our Chinese driver's preferred position was always on the left, in the wrong lane. If an oncoming car appeared, he would return most, but not all, of the way back to the right, and he was even less inclined to accommodate approaching motorcyclists, leaving them only about a quarter of a lane in which to slip past us. Every time we came to a curve, he would honk repeatedly to announce our approach. Later, on the major expressway leading to the central city, we joined a huge rush-hour traffic jam. All the cars had squeezed tightly together, transforming the four marked lanes into five, and whenever one of these improvised lanes seemed to be moving slightly faster than the others at least half the drivers on either side would try to force their way into it, while others drivers vied to squeeze into the spaces they were attempting to evacuate.

A couple of days later, the same driver picked me up at my hotel, and we drove across the city to pick up McGowan, who was living in an old quadrangle house, a *siheyuan*, in one of Beijing's *hutong*, which are ancient, densely settled neighborhoods transected by grids of narrow lanes and alleys. *Hutong* comes from a Mongolian word meaning "water well"; in Chinese today it signifies both a narrow urban alley and the neighborhood defined by it. McGowan's house was about a hundred yards up the alley, and I told the driver that if he would park at the curb on the main street I would run up and let McGowan know we had arrived. He laughed at this, and his meaning was easy to understand: We are in a car, so, obviously, we will drive! The alley in front of McGowan's house was wide enough for an automobile, but it ran one way in the wrong direction, and the driver, in order to reach the house by car, had to start a block away and make a long detour through other alleys, one of which was so narrow that he folded back the side mirrors on his Buick Regal to keep them from being snapped off. (Based on my own, nonscientific, observations, I would guess that there were more new Buicks in Beijing when I was there than there were in the United States.) As we inched along the alleys and squeezed around corners, we forced children, cats, bicyclists, and old women pushing carts to back up or press into doorways, to keep from being crushed, and the driver honked constantly to warn others to make way. No one seemed to be bothered by the noise or the inconvenience; the pedestrians apparently shared the driver's belief that the only rational thing to do with a car is to drive it. Our marauding Buick was public proof that we were people of significance, and the alley's residents treated us accordingly.

Alley neighborhoods like McGowan's are among the oldest in Beijing. The first ones arose in the thirteenth century in areas near the Forbidden City, and over succeeding centuries they grew to cover much of the central city. When I first found the McGowans' front door, in a tall, gray, windowless stucco-covered brick wall, I wondered whether I had come to the right place, because the doorway, though brightly painted, looked as though it might conceal a transformer or an incinerator. But the door led to an airy rectangular flagstone courtyard with rooms arranged around it. Beginning in the early twentieth century, old quadrangle houses like this one were often broken up into separate residences, room by room, and ramshackle additions and even complete dwellings were erected in their courtyards, a process that accelerated following 1949, when Mao (who had lived in a *siheyuan* not far from the McGowans') came to power. But the house in which the McGowans were living was still intact: two bedrooms, a kitchen, a dining room, a living room, and a family room, plus the courtyard, which had a couple of decent-sized trees growing in it. The McGowans were renting the house from a man who, they said, had been a government official during the Cultural Revolution; Rory's wife told me she shuddered to think what he must have done in his official capacity to have merited such a spacious place.

In Beijing's *hutong*, the buildings that line the alleys running east and west are mostly residential (giving their courtyard-facing windows maximum exposure to the south and the sun), while the buildings on the north-south alleys tend to be mainly commercial: a tiny store selling only cigarettes, another selling only vegetables, a closet-size butcher shop, a mechanic's work-bench. When my camera's AA batteries died, I bought replace-ments at an odds-and-ends shop the size of a packing crate. Not far from the McGowans' door, I walked past a grilled-meat res-taurant, whose customers were sitting in the alley itself, on single concrete blocks and old plastic antifreeze bottles. The chef stood at a hibachi-size coal or charcoal grill, which he had placed in a crude niche in a wall, and the customers ate from wooden skew-ers, which they dropped to the ground when they were finished. The skewers, along with the diners' cigarette butts and other refuse, would be cleaned up later by the neighborhood's bicycle-borne trash collectors, whose services cost each alley resident about a dollar and a half a year. Among the small businesses in the *hutong* were several tiny, brightly lit beauty parlors, each with a woman sitting in a chair, waiting for customers. Most of these women were actually prostitutes, McGowan told me, and could be distinguished from real beauticians by the fact that they would smile and nod and sometimes beckon when we looked in as we walked past, and by the fact that the rear portion of each shop had a floor-to-ceiling curtain drawn across it. "And no scissors," he added. Scattered through the alleys were fluorescent-lit public restrooms, mostly built within the past decade or so. (The McGowans' house was unusual in having its own toilet.) These were divided into men's and women's sections, but there were no stalls in the one I used one day—just a long urinal and round openings in the floor, for squatting. All day long, men and women on three-wheeled carts pedaled slowly past the Mc-Gowans' door, calling out whatever they had to offer. There were coal sellers, trash collectors, tinkers, recyclers, broom sellers, rice sellers, vinegar sellers, beer sellers, knife sharpeners. I saw an old woman sitting on an overturned bucket, making shoes by stitching uppers to soles, and an old man knitting very rapidly, and many old men looking after pet birds in hatbox-size bamboo cages. A giggling girl, on her way home from school, approached me and said, "Hello, how are you?"—trying out her English. I saw a woman washing clothes in a shallow bowl and then hanging them up to dry on one of the low, improvised-looking power lines that seemed to run everywhere. (The tangle of wires above one alley I wandered into looked like a bungled game of cat's cradle.) I stood for a long time with a group of men who were watching some other men building a brick wall. The masons were creating a compact living space for someone, a dirt-floored room about twelve or fifteen feet square, and around the corner from their work site I watched a half-dozen other men building another addition, using chisels, adzes, and other ancient-looking hand tools. Occasionally, if a door happened to open as I passed, I glimpsed a room with a bed jammed into it, sometimes illuminated by a TV. Everywhere, I saw people cooking and heating with smoldering cylinders of compressed coal, each somewhat larger than a jumbo can of tuna. Surplus coal cylinders were often stored outside, in stacks along the alley walls, as were residents' winter supplies of cabbages, leeks, onions, and carrots. I asked McGowan why these valuables weren't locked up indoors; he said there was no need, because neighbors (especially the elderly ones) watched out for one another, and pilferage, on the rare occasions when it did occur, was swiftly punished.[6]

There is an old Chinese saying that means, approximately, "There are 3,600 major *hutong* in Beijing while the minor ones are as numerous as the hairs on an ox." This is no longer mathematically or even metaphorically accurate, because in recent years the Chinese have been busy bulldozing the old neighborhoods, usually to make room for shopping malls, expressways, office towers, and blocks of grim, socialist-style apartment buildings, although sometimes with nothing more definite in mind than sweeping away the past. The Chinese don't necessarily share Western ideas about preserving old buildings (ideas that Americans actually haven't shared for all that long, either). McGowan told me that when the Chinese do decide to preserve an old building they tend either to tear down everything around it to create parking spaces for tourists, or to tear down the building itself and build a copy with modern materials. At the edge of one partially demolished old neighborhood, near one of Beijing's huge new ring roads, I saw a cluster of modern variations on old *hutong* buildings, with a long, narrow park running beside them. There was an elaborate photo display next to this development, featuring black-and-white photographs of the neighborhood that had been razed to make way for it. The point of the display was to document the blight that had triumphantly been eliminated.

Hutong residents don't always go quietly. Along an alley separating an old neighborhood from a new one, former residents of a demolished *hutong* had built a long row of new living spaces, lean-to style, against the outside wall of a bleak new apartment building. Some of these improvised dwellings were two stories tall, and most were made of crudely laid brick. I saw an old woman lowering a small bundle of garbage on a string down to a young girl, perhaps her granddaughter. On one low roof, cut-up pieces of onion were spread out to dry on a big double sheet of newspaper. Mostly, though, the old neighborhoods have simply disappeared.

Mourning the destruction of Beijing's *hutong* and the way

of life that they have sustained, in many cases for centuries, may just be Western meddlesomeness, the romanticizing of other people's poverty, or Boomer-style real estate envy. (If the *hutong* were in Manhattan, all the *siheyuan* would have been meticulously restored by yuppie lawyers and investment bankers, and would now be the most coveted apartments in the city.) Living conditions in many of Beijing's vanished *hutong* were squalid even by local standards, and many former residents undoubtedly view their new circumstances, however uninspiring they may seem to a naïve Westerner, as a providential upgrade. But it's hard not to think that the *hutong* represented, and still represent, a potentially valuable model for how to sustain a large population with minimal dependence on motor vehicles. A functioning *hutong* embodies nearly all the qualities that Jane Jacobs found central to urban life—the compactness, the productive jumble of thriving uses, the deep networks of personal interconnection, the reduced reliance on motorized transportation. Rory McGowan's alley neighborhood was nowhere near as dense as Manhattan or Hong Kong, but its energy footprint was extremely small, and the community it enclosed was vibrant and self-supporting. When the bicycle of one of the McGowans' sons broke, the local repairman set up shop outside the McGowans' front door, and the boys chatted with him while he worked.

A thriving *hutong* is a model of efficiency, a tightly interdependent human ecosystem. The American journalist Peter Hessler, who lived in Beijing for several years in a modern apartment building adjacent to an alley that was so small it didn't have a name, described his neighborhood in an article in *The New Yorker* in 2006:

> On an average day, a recycler passes through every half hour, riding a flat-bed tricycle. They purchase cardboard, paper, Styrofoam, and broken appliances. They buy old books by the kilogram and dead televisions by the square inch. Appliances can be repaired or stripped for parts, and the paper and plastic are sold to recycling centers for the barest of profits: the margins of trash. Not long ago, I piled some useless possessions in the entryway of my apartment and invited each passing recycler inside to see what everything was worth. A stack of old magazines sold for sixty-two cents; a burned-out computer cord went for a nickel. Two broken lamps were seven cents, total. A worn-out pair of shoes: twelve cents. Two broken Palm Pilots: thirty-seven cents. I gave one man a marked-up manuscript of the book I'd been writing, and he pulled out a scale, weighed the pages, and paid me fifteen cents.[7]

Refuse that can't be recycled is left on the ground, outside the front door, for collection by other tricycle carts. Such trash piles are usually tiny—a handful of cucumber peelings, a few eggshells, a couple of burned-out coal cylinders—because anything of conceivable value has already been reused or set aside for the recyclers, and because *hutong* commerce employs very little disposable packaging.

The compact symmetry of *hutong* life doesn't carry over into the new Beijing. Most of the recently constructed residential areas in the central city are seemingly denser than the *hutong* they have replaced, because the tall apartment buildings house more people on smaller sites than the old alley neighborhoods did; but the fertile, efficient hodgepodge of uses is greatly reduced, and Beijing's minimal public-transit system doesn't come close to being an adequate replacement for the old pedestrian-and-bicycle networks of the alleys. Beijing is currently expanding its subway system, but that move is almost an afterthought; far more effort and investment have gone into creating vast new ring roads, expressways, and parking lots. The contrast between the old and new cities is even more apparent in the entirely car-dependent subdivisions that make up Beijing's suburbs, where many foreigners and newly affluent Chinese now live, essentially in self-exile. On the top floor of Beijing's strangely fascinating Urban Planning Museum, near Tiananmen Square, you can get a sense of the city's new proportions by walking around on top of a huge photographic city map, which was created from satellite images and three-dimensional building models. I visited the museum with an American friend who was living in a gated expatriate community far from the city center—a subdivision that she said reminded her of Seahaven, the phony town in the movie *The Truman Show*—and we walked over the map along the route from my hotel to her house, on the expressway that leads to the airport. From above, the new Beijing looks disconcertingly like Atlanta, another rapidly growing city whose economic and environmental destinies have been shaped primarily by cars. The Chinese would be better off if they looked else-where, or inward, for inspiration.

China and many other non-Western countries are rapidly urbanizing. That is, their populations are undergoing a major migration from rural areas to cities. This trend, which has been under way all over the world for decades, is often derided by American environmentalists, who have generally preferred to see people moving in the opposite direction, toward "the land." But urbanization is usually a good thing, both for the direct partici-pants and for civilization in general. City families live more compactly, do less damage to fragile ecosystems, burn less fuel, build stronger social ties to larger numbers of people, and, most significantly, produce fewer children, since large families have less economic utility in dense urban settings than they do in marginal agricultural areas. Wealthy westerners are capable of romanticizing truly desolate urban living conditions, as was evident with Hollywood's embrace of the 2008 movie *Slumdog Millionaire*. But humanely managing wholesale urban migration will play a critical part in any quest for global sustainability. Stewart Brand has written, "Already, as a result of headlong ur-banization, birthrates have plummeted in the developing world from 6 children per woman in the 1970s to 2.9 now. Twenty 'less developed' countries, including China, Chile, Thailand, and Iran, have already dropped below the replacement rate of 2.1 children per woman."[8] That's good news, both locally and globally, and Brand sees hope for the world in the spread of dense, jumbled urban neighborhoods very much like the ones that the Chinese have assiduously been leveling in Beijing. In addition, recent studies have linked urbanization to the rapid reforestation of areas that had been cut for timber or cleared for agriculture. "By one estimate," Elizabeth Rosenthal wrote in *The New York Times* in 2009, "for every acre of rain forest cut down each year, more than 50 acres of new forest are growing in the tropics on land that was once farmed, logged or ravaged by natural disaster."[9]

When American environmentalists contemplate China's future, they tend to focus on high-minded utopian schemes, like the creation, from scratch, of entire new communities—such as Dongtan Eco-City, a high-tech "carbon neutral" development planned for an island in the mouth of the Yangtze River, near Shanghai—and on American-style defensive preservation, as in the cordoning off of isolated, relatively undefiled natural areas, to protect them from encroaching development. But these two strategies are really just manifestations of, respectively, LEED brain and what might be thought of as Nature Conservancy brain, and they don't truly move China or the world closer to a real accommodation with the future. Dongtan Eco-City represents a huge investment of cash, modern technology, and Western engineering, yet even if it fulfills its developers' ultimate goal, for a city of 500,000 by 2050 (on a marshy island three-quarters the size of Manhattan), China's main need is not for isolated "demonstrator" suburbs intended for relatively small groups of relatively wealthy commuters—or for scattered pockets of what American urban planners usually refer to as open space. What China and the rest of the world need is a way of arranging large populations compactly and efficiently, without an overwhelming dependence on automobiles. Manhattan, Hong Kong, and some of the older sections of Beijing offer far more instructive examples of how to achieve low-impact urban living, and they exist right now. They are better models for the future than the city of expressways and automobile commuters into which Beijing is rapidly transforming itself, and are better, even, than Dongtan Eco-City.

SIX MONTHS BEFORE MY TRIP TO BEIJING, I VISITED another city that has often been in the news in the West in recent years: Dubai, in the United Arab Emirates. Dubai is far smaller than Beijing, but in many ways the example it sets is more disturbing, because it has grown from virtually nothing within just the past couple of decades, and because its planners, from the beginning, have seemed determined to create an environmental disaster.

Before traveling to Dubai, I hadn't really known what or where it was. Now I can tell you that it's both a city and a state, and that it's on the southeastern spur of the Arabian Peninsula, east of southern Saudi Arabia and northwest of Oman. A nonstop flight from New York City lasts almost thirteen hours, and crosses Iran. (The route map shown on the video screens in the cabin during the flight looks like a graphic from a CNN story about trouble in the Middle East.) Dubai is the second-largest, in terms of land area, of the seven semi-independent hereditary sheikhdoms that constitute the UAE; the largest of those sheikhdoms is Abu Dhabi, whose central city is about eighty miles from Dubai's.

Oil was discovered in Dubai in the 1960s, but it directly contributes only a small percentage of the emirate's current income; the economy's main elements are tourism, frenetic real estate development, and no-questions-asked international commerce. (I spent one afternoon with a French-Lebanese man named Fadel, who told me that he had lived in Dubai for two years and that he was in the business of importing uncut diamonds from Africa.) Dubai's tourists come from northern Europe, the United Kingdom, other Arab countries, and Asia, among other places. Shortly after my visit, my wife's brother, who was living in Moscow at the time, took his (Russian) wife and their two young children to Dubai for a two-week vacation. "It's sort of the latest place for Russians to go, after Turkey, Thailand, Greece, etc.," he told me, by e-mail, from his beachfront hotel. "It has a little more cachet than those other places, but our hotel is still less than two hundred dollars a night. We came here because we wanted the kids to be able to swim every day, and in places like Greece and Cyprus the water is still relatively cold. Here, it's like swimming in your bathtub." Dubai gets so hot, in fact, that the water in many hotel pools is not heated but chilled, to keep swimmers from feeling poached.

Dubai has grown so fast that no one knows what its population is. I heard a million, I heard two million, I heard whatever. The number of "nationals," or native Emiratis, is just 200,000 or so, a small minority. At least half the people in the emirate are laborers and low-level service workers from India, Pakistan, and Bangladesh. They toil long hours in brutal conditions for minimal wages, live ten to a room in company-owned labor camps, often ship their earnings home in the form of miniature gold bars, and must leave the country when they can no longer work. Construction continues all night, under lights, and is seemingly going on everywhere, all the time, all at once. The most ambitious project that was under way when I was there was a skyscraper called the Burj Dubai, which, though it's still incomplete at the time I'm writing, is already the tallest man-made structure in the world. (*Burj* is Arabic for "tower.") When it's finished, it will be almost exactly a half-mile tall, or about twice as tall as the Empire State Building, and will serve as the centerpiece of a vast commercial and residential development, called Burj Dubai Downtown, which will eventually include many dozens of lesser high-rises, an artificial lake, the world's largest shopping mall, and numerous other attractions. The Burj Dubai, like all buildings in the region, is literally built on sand: bedrock lies too far below the desert's surface to serve as an anchor for a foundation.

When the first version of Dubai International Airport was built, in 1960, it had a single runway, which was made of compacted sand. Now landing in Dubai feels like landing at LAX— arriving planes descend through layers of phlegm-colored automobile exhaust—and the state-owned airline, Emirates, is one of the busiest in the world. The government is currently not only expanding Dubai International but also building a second huge airport, Al Maktoum International, which will be the size of Heathrow and O'Hare combined and will be surrounded by hundreds of commercial and residential high-rises. Change in Dubai has occurred so rapidly that it can't be charted on an ordinary timescale. In 2008, an American travel writer said that Dubai "is growing so fast that its newest developments can only be measured in hummingbird flaps."[10] I played a round of golf at Emirates Golf Club with a group that included Maurice Perry, who is the director of development design for Jumeirah, a worldwide builder and operator of luxury hotels, among them Dubai's famous Burj Al Arab, the one that looks like an enormous sailboat about to ground itself at the edge of the Persian Gulf. Perry was born in Australia in 1945, and has lived all over the Far East. ("Have you played golf in the Philippines? You must!") He has a trim salt-and-pepper beard, and on the day we met he was wearing a golf shirt with the eagle-and-lion logo of the Singapore Island Country Club, to which he once belonged. I asked him where he was living now.

"Here in Dubai," he said.

"And how long have you lived here?"

"Six weeks."

That made him almost an old-timer. Perry's apartment was in a three-year-old building in a section of man-made waterfront called Dubai Marina, which will eventually contain more than two hundred high-rises; it was desert at the turn of the millennium. I met British expatriates in Dubai who talked about the early 1990s as though that were a vanished golden age when everyone knew everyone else and hung around in the same restaurants after work. Occasionally, I met someone who had been in Dubai even longer, all the way back to the seventies or eighties, when the emirate was just a bump in the desert, known for its pearl diving. Meeting someone like that was like running into a New Yorker who could remember buying Manhattan from the Indians. Photographs of Dubai from fifty years ago look like something out of *Lawrence of Arabia*: a few small structures here and there, surrounded by miles of blowing sand.

Viewed from the air, many parts of Dubai look as dense as Manhattan or Hong Kong, but the city is actually even more automobile-dependent than Beijing is. In fact, it's more automobile-dependent than any city in the United States. This is doubly lamentable because modern Dubai was created on what was essentially a blank sheet of paper, and was conceived at a time when anticipating the long-term futility of building a society based on cars didn't require extraordinary foresight. One of the frustrations of urban planners in New York City is the impossibility of, say, rolling up Lexington Avenue in order to repair, modernize, and rearrange the underlying subway tunnels, water mains, power lines, and other aging infrastructure. (Subterranean workers in Manhattan still sometimes encounter century-plus-old water pipes and electrical conduits made of wood.) Dubai's builders, by contrast, began with empty desert—

yet they failed to make almost any provision for the future, or to think through how the city might function once the frenzy of construction and real estate speculation had slowed. (Never mind the rationality of deciding to build at all.) Until very recently, Dubai had no true public transit, other than the stifling buses that carry imported workers between their job sites and their camps. The city, belatedly, is building an ultramodern, fully automated subway and rail system, called Dubai Metro, but this attempt to graft transit onto a city like Dubai, even though the project is backed by what occasionally appears to be all the money in the world, is an enterprise destined to disappoint. People who use the new trains will still face the challenge of getting themselves from their metro stop to their final destination, since Dubai must be one of the least walkable cities in the world. I stayed in one small hotel and two big ones—including the Burj Al Arab—and there was no plausible destination to which I could have traveled on foot from any of them. Going from virtually anywhere in Dubai to virtually anywhere else means getting into a car and plunging into the permanent traffic jam that hogties the central city. Traffic during my visit was snarled not only by the huge number of cars but also by Dubai's Sisyphean efforts to make room for still more drivers: existing expressways and interchanges, whose concrete had scarcely had time to harden, were being torn apart and widened, necessitating detours and delays. It is illegal to drive in Dubai with any alcohol whatsoever in your bloodstream, and the penalties for violators are severe: heavy fines, mandatory prison terms, deportation. Even so, Dubai has one of the highest traffic-fatality rates in the world, mainly because the roads are as crowded as L.A.'s freeways at rush hour, and the taxi drivers and the rich young nationals are legendarily reckless. (A popular local pastime is "dune bashing," or driving at high speed through the sand on the city's outskirts.)

Before traveling to Dubai, I had imagined it as very hot and very dry, like southern Arizona at noon. I got the hot part right, but I was unprepared for the humidity. The Persian Gulf acts like a vast steam bath, and in summer the scalding dampness combines with automobile exhaust and windblown desert dust to create a sort of suffocating fog, which, on bad days, feels unsafe to stand in, much less to breathe. Yet the city's growth has been explosive. The Burj Al Arab is the tallest hotel in the world, and it stands more than a thousand feet from shore, on its own tiny island, partly to discourage sightseers (who aren't allowed to cross the guarded causeway) and partly, I was told, to keep the building from casting shadows on sunbathers. The view from the floor-to-ceiling windows on both floors of my two-story suite felt like the view from an airplane, or the space shuttle: I looked down, though heavy mist and fumes, toward the World, a Mercator-inspired arrangement of artificial islands, then under construction. Those islands, along with three similar artificially created offshore residential developments, are some of the most expensive real estate on earth. Day and night, convoys of trucks arrive in the city from far away, carrying loads of imported boulders, for the landfill on which the developments stand.

One persistent mystery in Dubai has to do with what the people who buy real estate actually own. Only nationals can own property in Dubai, except in a few designated zones, and even there the issue is ambiguous. Virtually all real estate deals with foreigners are actually ninety-nine-year leases—although that doesn't seem to be a disincentive for the kinds of people who roam the world with suitcases full of hard currency. In one residential development I visited, the largest house had twenty-five bedrooms, comprised 70,000 square feet of interior space, and was owned by the king of Swaziland (where the gross domestic product per capita is about $5,000 and the average life expectancy at birth is 33.2 years); the second-largest house in the development, I was told, was owned by a Nigerian, occupation unknown, perhaps something to do with guns. A resident of the development told me that a friend of his had bought a house not far away, a year before, for several million dollars and had been moving his furniture into it when a man approached him and asked him what he would take for the place. He named what he felt was a ridiculous price; the buyer offered cash, and he moved his furniture back out. Another resident told me that he had agreed to buy a condominium in a new development but didn't yet know what his new address would actually be because the emirate's ruling sheikh, who is the ultimate owner of everything and therefore controls all real estate decisions, had decided to change the location of the entire development.

Dubai's main tourist event of the year is the month-long Dubai Shopping Festival, which takes place in early winter and features deep discounts at retail stores throughout the city. The festival has been so successful that the city created a second shopping festival, called Dubai Summer Surprises, in the hope of generating more trade at a time of the year when outdoor temperatures sometimes top 125 degrees. One evening, I took a taxi to the Mall of the Emirates, which at the time was the largest shopping mall in the city (it has since been superseded) and one of the largest outside the United States. On the way there, my cab passed a row of car dealerships, which in Dubai are enormous: daytime temperatures are so high that cars can't be displayed outside, so the dealerships all have vast, air-conditioned showrooms, with additional inventory stored underground. When I arrived at the mall, at around eight-thirty, a large crowd had gathered in the central atrium. I rode two escalators to the third level, found a small gap in the rows of people lining the balcony railings, and looked down to the ground level, where I saw the reason for the excitement: a stage, a drum kit, a runway, and a big sign that said, "Tommy Hilfiger Fashion Show and Personal Appearance." The show—imported models wearing Hilfiger's latest preppie-inspired creations—began a little later. When it was over, Hilfiger himself appeared, looking like a natty member of the forty-third form. He was making his first visit to Dubai, to open the first of twenty-five Hilfiger stores in the Gulf region, part of an ongoing wave of Western investment.

The Mall of the Emirates looked like every mall you've ever seen, with lots of the same stores: Ralph Lauren, Bombay & Company, Harvey Nichols, Yves St. Laurent, a few hundred others. (Gap and Banana Republic were coming.) The food choices were familiar, too: McDonald's, KFC, Subway, Dairy Queen, Johnny Rockets. My Russian sister-in-law told me that prices at the malls in Dubai were lower than in Europe but higher than in the United States—important considerations for someone from a country that has yet to develop the concept of the anchor store. A little farther down the corridor I passed a discount underwear shop, inside which I saw two Muslim women wearing black abayas with full head-and-face covering, shopping for bras. At the far end of the same long corridor I came to Ski Dubai, the city's famous indoor ski slope—which a 2006 article in the travel section of *The New York Times* called "Dubai's marquee attraction."[11] Actually, if you've ever seen snow and don't work in the refrigeration business, Ski Dubai is unlikely to impress you. The total run is just 1,200 feet, and the ambience is less Aspen than Discovery Zone, especially in the dank changing area, where customers are fitted for skis and identical red-and-blue insulated coveralls. Ski Dubai is a lot like the rest of Dubai: you're amazed that anyone had the money and the moxie to pull it off, but after you've gotten past the novelty you can't help wondering whether people haven't simply lost their minds. One of the many mega-projects that were under way during my visit was an amusement park and residential development called Dubailand, which has been under construction since 2003 and still has a few more years to go. When it's completed, it will be bigger than Disneyland and Disney World combined. I passed the site on my way to a condominium-and-golf-course development at what was then the far edge of the city, and saw mainly roads and parking lots and glare and blowing sand.

Oil revenues account for just 6 percent of Dubai's income at the moment, and that income stream will decline in coming years, as Dubai's (and the world's) remaining oil reserves are drawn down. Dubai's ruler, Sheikh Mohammed bin Rashid Al Maktoum, has been praised by many for building a Middle Eastern economy that doesn't depend on oil—but in actuality Dubai is at least as oil-dependent as any other prosperous country in the world. Its economy is based on jet fuel, gasoline, unlimited air-conditioning, and the transmutation of seawater into hotel showers (all the fresh water in Dubai is created in desalination plants on the Persian Gulf), and as the global oil market exhausts itself Dubai will feel the consequences more painfully than even Saudi Arabia or the United States. Dubai has often called itself a city of the future, but it is actually a city of the past, an explosively growing monument to unsustainability.

Dubai's example is even more discouraging than Beijing's, because in Dubai's case—given the decision to build it at all—there were no financial or infrastructural constraints on the city's evolution. Out of all the possible development models available in the world, Dubai's rulers somehow seem to have settled on Las Vegas—a city that, by comparison, feels almost homey, even eco-friendly. Some visitors to Dubai have commented that parts of the city's skyline remind them of New York's, but the resemblance, to the extent that it exists, is illusory. There are many tall buildings in Dubai, as there are in Manhattan, but they are disconnected points in a sprawling array whose true designers were the automobile and the jumbo jet. Dubai is a doomed city, destined to collapse under the weight of its energy needs.

WHEN I TOLD A FRIEND THAT I THOUGHT NEW YORK City should be considered the greenest community in the United States and a useful model for the future, she looked puzzled, then asked, "Is it because they've started recycling again?" Her question reflected a central failure of the American environmental movement: too many of us have been made to believe that the most important thing we can do to save the earth and ourselves is to remember each week to set our cans, bottles, and newspapers on the curb. Household recycling is popular because it enables us to relieve our gathering anxieties about the future without truly altering the way we live. But most current American recycling has, at best, a neutral effect on the environment, and some of it is demonstrably harmful. As William McDonough and Michael Braungart point out in *Cradle to Cradle: Remaking the Way We Make Things*, which was published in 2002, most of the materials we place on our curbs are merely "downcycled"— converted to a lower use, providing a pause in their inevitable journey to a landfill or an incinerator—often with a release of toxins and a net loss of energy, among other undesirable effects, especially if the recycled materials traveled to their collection center in the back of someone's SUV.[12]

Recycling makes people feel good. "I get a warm and fuzzy

feeling when I fill a recycling bin every week," a participant in an online forum wrote not long ago. We all know that feeling, which, paradoxically, can make generating trash feel like a moral act: the bigger the bottle pile, the better. On vacation a few years ago, I was with a group of adults when they learned that a local collection program for recyclables had been suspended and that there was no point in separating their newspapers from the regular trash. They were appalled. One of them said that putting a newspaper in a wastebasket now felt evil to him, like throwing trash onto the ground. Yet he hadn't expressed even a twinge of guilt about jumping into his huge car that morning to make the nine-mile round-trip to the nearest store to pick up the very same newspaper, along with a cup of coffee to go. Building public aversion to heedless waste is a good thing, but allowing people to believe that dividing their trash into two piles is an adequate response to the world's present problems is not. Intelligent recycling will become increasingly important, but our newspapers, bottles, and cans are very small elements in a large and distressingly complex environmental puzzle.

When affluent Americans think about "going green," they tend to focus on enhancements to their own consumption rather than on subtractions from it: buying a new, more fuel-efficient car (rather than driving less or taking the bus), building a new kitchen full of eco-friendly gadgets and exotic building materials (rather than deciding not to add yet another underused room to their house), replacing their old windows with high-tech new ones (rather than caulking air leaks, drawing the curtains during the day, and turning the air-conditioning down or off), and eating better-tasting chickens, tomatoes, and eggs. Perhaps this way of thinking is unavoidable in the most ravenous consumer society in the history of the human race. Still, our ability to make self-serving rationalizations can be breathtaking. A wealthy young family recently remodeled and hugely enlarged a big house not far from where I live. The husband's job is connected to some form of environmentally oriented construction, and he initially intended for the renovation to include the installation of both a wind-powered generator and a geothermal heating-and-cooling system—mainly as instructive examples for others, since neither would represent a rational return on investment. Both ideas turned out to be financially senseless, even for him, to his regret. Yet he apparently felt no hesitation about greatly increasing the size of what was already a very large house. The renovated house, undoubtedly, contains all the usual eco-features. But the greenest choice, under the circumstances—making do with (or reducing) the existing building envelope, while improving its energy efficiency—was apparently out of the question.

In 2008, *The New York Times* ran an article in its travel section which neatly captured this same kind of acrobatic self-justification. The article's author, Jennifer Conlin, set out to see if she and a friend could travel from London to Australia "in a more sustainable way than I have grown accustomed to over the years." She "offset" her share of the carbon footprint of her thirty-eight-hour transcontinental round-trip flight by making a $21.50 contribution, through Qantas, to an Australian "greenhouse gas abatement" program, and took comfort from the words of the founder of the British website responsibletravel.com, who told her that international travel "keeps many cultures from going extinct. Often rituals and traditions are passed down between generations primarily because tourists come to see them." She also booked a room at a self-styled "eco-lodge," drank only Australian wines while in Australia, traveled by train when possible, and ate kangaroo.[13]

Conlin's particular form of rationalization should be familiar to most of us. It's a little like agreeing that your doctor was right when he urged you to get more exercise, but then—instead of simply jumping on your bike or going to the gym—analyzing your current activities and discovering that you've been burning more calories than you would have guessed while doing things like sleeping, showering, and sitting at your desk, and then deciding that you don't need to make any lifestyle changes after all, because, it turns out, you've been getting more exercise all along. (So-called carbon-offset schemes are especially suspect, since they are highly vulnerable to exaggeration, self-deception, and fraud.) I recently heard a young couple debating whether it was environmentally more responsible to drink wine from California (transported across the United States by truck) or from France (transported across the Atlantic Ocean by ship). This is exactly the sort of environmental dilemma we all like to find ourselves in: Which is better for the earth, the Hawk Crest or the Pouilly-Fuissé?

Similar issues are raised by the current popularity, among

certain caring, affluent consumers, of favoring locally grown food—so-called locavorism. I, too, greatly prefer the produce I purchase at farm stands and farmers' markets, in comparison with the stuff that's usually available at my grocery store, but the idea that such a preference is in any sense "sustainable" depends on arithmetical sleight of hand. The distance that a particular food item travels between its grower and its ultimate consumer is not an accurate measure of the amount of energy that was required to put it on the table; far more significant factors are: how it was grown, how it got where it was going, and what else was traveling with it. The California raspberries I purchase at my grocery store have a smaller carbon footprint than the local raspberries I picked recently at a farm just a couple of towns away, because the California raspberries crossed the country in a shipment containing tons of other produce, and therefore represent a minute expenditure of fuel per berry, while the local raspberries were obtained by my wife and me during a thirty-mile round-trip in a car whose only other cargo was ourselves. There is no sense in which my preference for local raspberries represents a gain for the environment, since the energy expenditure and carbon output per unit of food were vastly higher for the local berries than they were for the ones originating 3,000 miles away. The writer Michael Pollan has been a passionate and effective advocate for small-scale, locally oriented agriculture, but many of the mouthwatering examples he has written about undercut the increasingly common carbon argument for preferring such food. In his book *The Omnivore's Dilemma*, published in 2006, he describes an extraordinarily appealing organic farm, whose owner views even Whole Foods as ominously industrial. Among the farm's regular customers are some people who have taken "a beautiful drive in the country" in order to resupply themselves—including one who explains, "I drive 150 miles one way in order to get clean meat for my family."[14] This is a powerful endorsement, and it made me want to eat those chickens, too, but there is nothing remotely sustainable about three-hundred-mile car trips to buy dinner. All the car miles traveled by the farm's loyal customers should be considered part of the carbon footprint and embodied energy load of those chickens.

The weakness of locavorism as a global environmental principle is easy to see if you extend it beyond the kitchen. *The Omnivore's Dilemma* has sold many thousands of copies all over the world, and those copies have been shipped at least as many miles as the food on the shelves in your grocery store (and usually in far smaller batches, using less efficient conveyances), yet no author, Pollan included, would argue that readers should buy only books produced within a few dozen miles of where they live. Someday, perhaps, all our reading matter will be delivered to us electronically; until then, should environmentalists write only for their neighbors? Likewise with clothes, computers, appliances, low-flow plumbing fixtures, oil, and everything else we consume. (How are the locally produced hybrid cars in your neighborhood? And what about locally grown coffee beans in Portland and Seattle?) Global trade can actually reduce carbon output, by concentrating production in the places where production is the most efficient and by eliminating redundant, inefficient car trips taken by individual shoppers. (The same people who make an environmental case for eschewing food grown in distant places often make the opposite case for ordering other goods online.) Pollan's books constitute a passionate, persuasive argument for a radical transformation of American agriculture and for the elimination of the ill-considered government subsidies that have helped to turn many of the most common American foodstuffs into nutritional simulacra, but nothing could be less sustainable than a world whose 6 billion (and counting) residents had to rely totally, or even mainly, on locally produced anything. Shipping foodstuffs and other goods long distances—from areas of abundance to areas of need—will become more important, not less, as the world's energy and emissions difficulties deepen.

In 2008, Michael Specter addressed the issue of "food miles" in a *New Yorker* article on carbon footprints; in it he quoted, among other authorities, Adrian Williams, a British agricultural researcher, who told Specter, "The idea that a product travels a certain distance and is therefore worse than one you raised nearby—well, it's just idiotic. It doesn't take into consideration the land use, the type of transportation, the weather, or even the season. . . . This is not an equation like the number of calories or even the cost of a product. There is no one number that works." Specter wrote:

The environmental burden imposed by importing apples from New Zealand to Northern Europe or New York can be lower than if the apples were raised fifty miles away. "In New Zealand, they have more sunshine than in the U.K., which helps productivity," Williams explained. That means the yield of New Zealand apples far exceeds the yield of those grown in northern climates, so the energy required for farmers to grow the crop is correspondingly lower. It also helps that the electricity in New Zealand is mostly generated by renewable sources, none of which emit large amounts of CO_2. Researchers at Lincoln University, in Christchurch, found that lamb raised in New Zealand and shipped eleven thousand miles by boat to England produced six hundred and eighty-eight kilograms of carbon-dioxide emissions per ton, about a fourth the amount produced by British lamb. In part, that is because pastures in New Zealand need far less fertilizer than most grazing land in Britain (or in many parts of the United States). Similarly, importing beans from Uganda or Kenya—where the farms are small, tractor use is limited, and the fertilizer is almost always manure—tends to be more efficient than growing beans in Europe, with its reliance on energy-dependent irrigation systems.[15]

In 2007, the huge British supermarket chain Tesco announced a plan to favor locally produced foods and to label every product it sells with a calculation of its overall carbon footprint, including the cumulative footprints of its ingredients or parts, enabling consumers to measure the environmental impact of their purchases "as easily as they can currently compare their price or their nutritional profile."[16] The plan has been applauded by environmentalists and thoughtful customers, but the program, if it's ever fully implemented, can only be misleading, for the reasons that Williams enumerated. The overall environmental impact of almost any food item or other product depends on far too many complex variables—which themselves can change rapidly, and are subject to dispute and to changes in scientific opinion, as well as to creative fudging by manufacturers—to be conveyed by a grocery store carbon label. (A better path to the same goal might be to intelligently tax energy consumption and emissions, thereby automatically giving every product a simple indicator of its overall environmental impact: its price.)

Locavorism is appealing as an environmental strategy because it permits its practitioners to believe they're doing good for the world by doing well for themselves, and to recast their own consumption and nutrition preferences as contributions to humanity—like that "green" trip to Australia taken by the reporter for the *Times.* The fascination with locally produced food is also responsible for the absurd idea that it makes environmental or economic sense for city dwellers to become better locavores by growing their own crops. Dickson Despommier, who is a professor of public health at Columbia University, in New York, has promoted what he calls "vertical farms"—thirty-story urban towers in which produce would be raised hydroponically, for consumption by nearby city residents. Despommier claims many advantages for growing food inside skyscrapers: year-round crop production, isolation from the vagaries of weather, elimination of the need for pesticides and fertilizers, numerous others.[17] But building high-rise farms in dense urban cores would make no more sense than moving isolated apartment towers into Kansas wheat fields. Any conceivable benefit to be gained by shrinking the distance between the production and the consumption of particular food items would be more than negated by the inefficiencies inherent in increasing the distance between other human uses, and by the necessity of creating and maintaining the wasteful infrastructure needed to support them. If farming in skyscrapers makes agricultural sense, then, by all means, let's do it—but not in places where the environment would be served better by stacking people rather than crops. Vertical farming in Manhattan would make the city less green, not more. You can't save the world by trying to make dense urban areas more like the country. On the contrary.

THE CURRENT PUBLIC FOCUS ON ATMOSPHERIC CARBON— which in recent years has dominated discussions about the environment—is tricky for most of us because the likely global benefits of reducing the world's production of greenhouse gases are impossible both to quantify with certainty and to make tangible to those being called upon to sacrifice. Halting the dumping of toxins into a river leads to an immediate improvement in water quality and, therefore, to an immediate improvement in the lives of those who depend on the river as a resource; reducing the world's production of carbon dioxide and other harmful gases may slow the rate at which various catastrophic climate changes are likely to occur, but it will not reverse the changes already under way or, in itself, immediately improve the daily lives of those who bear the cost of making the reductions. (Even if civilization disappeared tomorrow, the environmental effects of global warming caused by humans would continue for decades.) The fact that the main likely beneficiaries have yet to be born makes it difficult not only to reckon the present value of actions taken in their behalf but also to assess the ultimate effectiveness of whatever actions might actually be taken, and it leads to the public-policy equivalent of playground arguments— "My father's carbon footprint is smaller than your father's"—and to politics-driven initiatives of questionable value.

Actually, there's a potentially productive way to think about carbon dioxide and climate change which doesn't depend solely on civilization's willingness to engage in global-scale delayed gratification, and doesn't depend even on achieving a worldwide consensus about causes and effects. Almost all human activities with large carbon footprints are going to become increasingly expensive and untenable for reasons that have nothing to do with their likely impact on the earth's climate fifty or a hundred years from now and can therefore be addressed with tools that don't depend solely on hypothetical arguments about the future, or on moralizing by environmentalists. Air travel is a significant contributor to atmospheric levels of greenhouse gases. That fact alone is unlikely to dissuade present-day travelers from hopping the globe at will—one of the most passionate advocates of carbon-reduction I know does his own flying on a Gulfstream V—but uninhibited air travel also has more immediate environmental, economic, and personal consequences, which we are in a better position to assess and respond to because they affect our lives directly, and right now. In 2007 and 2008, as the rising price of jet fuel made flying increasingly expensive and unpleasant, the appeal of casual globetrotting fell, and flying's contribution to individual carbon footprints fell with it, however slightly. That trend was interrupted in late 2008, when the price of oil plunged; it will resume when jet fuel again becomes prohibitively expensive, causing airlines to shrink their schedules further, and forcing financially strapped travelers of all kinds to reassess the urgency of trips that, ten years ago, they would have taken without a second thought. The actual environmental benefit in 2008 was small, but the trend, while it lasted, was pointing in the right direction, and it didn't depend on what anybody thought about carbon dioxide. It also suggests a general strategy, which is to address long-term but hard-to-define environmental problems by exploiting economic trends whose impact on living people is immediate and direct. This, incidentally, is a staple of self-help instruction: people are generally more successful at making long-term lifestyle changes if they focus on short-term consequences, even if those short-term consequences are actually secondary, in the larger scheme of things. Passionate lectures about lung cancer are usually less effective at persuading young smokers to quit than a complaint by a boyfriend or girlfriend about the unpleasant taste of a kiss.

Large numbers of environmentalists, scientists, economists, small-car drivers, and others have been condemning Americans' love affair with oversized automobiles since the early 1970s, yet even as those critics were becoming more adamant, the same cars were steadily growing larger. Car manufacturers claimed they couldn't survive without "light trucks," and Congress never found the moral strength to raise mileage standards or to eliminate the regulatory quirks that sustained the economic appeal of SUVs. No progress seemed conceivable, even though SUV drivers themselves routinely said they would be willing to drive smaller cars if other SUV drivers would switch, too. Yet all it took, finally, to transform the U.S. automobile industry was a single encounter, in 2008, with four-dollar-a-gallon gasoline.

The resulting changes to the automobile industry do not, in any sense, represent a "solution" to our ongoing petroleum crisis; the actual decline in both driving and fuel consumption has been small, and our society remains as automobile-dependent as ever. But the changes that were put into motion by more expensive gasoline did constitute a far larger and faster restructuring of the American automobile industry than could possibly have been achieved through continued public sermonizing about atmospheric degradation or, even, through a genuine tightening of federal fuel-efficiency standards. Money really does talk—and this fact suggests another powerful environmental tool, if we can find the will to use it: people do what they are given incentives to do.

Unfortunately, many critical structural incentives in the United States actually:

- guarantee harmful outcomes (as was also the case in the buildup to the present credit crisis, in which lenders and borrowers were extravagantly rewarded for taking the steps that led to their self-destruction); or
- cater primarily to political and commercial self-interest (as in the promotion of ethanol, which has made politicians look busy on the energy issue and has enriched the producers of ethanol but has harmed just about everyone else); or
- make people feel better without accomplishing anything substantive (as in the economic abracadabra that makes homeowners feel they're saving the world by installing heavily subsidized solar panels on the roofs of suburban McMansions).

We have encouraged sprawl, automobile dependence, and energy gluttony from the beginning, in part by concealing from ourselves the true long-term and short-term costs—and when potentially useful disincentives have arisen spontaneously we have generally rushed to neutralize them, as in the demands, in late 2008, that fuel taxes be reduced to bring down the price of gasoline, or in work-relief proposals to help revive the economy by building new roads or making it easier for people to buy new cars and build new houses.

But the power of well-designed incentives and disincentives is not in doubt. If there is a plausible way to reduce energy use and greenhouse-gas production in prosperous countries it almost certainly involves some form of taxation on fuels, a strategy that has the additional benefit of pushing us away from dependence on energy sources of which the world possesses a limited supply. One way to do that would be to maintain the price of energy at artificially high levels, and to eliminate the many structural incentives that lead to the creation of inherently wasteful communities. We know that high prices generate efficiencies, by creating permanent inducements to live smaller, live closer, and drive less. The question is whether we have the necessary foresight and courage to impose them on ourselves—and whether we are willing to endure the personal and national economic consequences of consuming less, especially right now.

In 2008, I attended a seminar on energy at a scientific conference in New York City. One side effect of attending such events is the feeling of despair that inevitably comes from hearing well-informed people speak about global environmental problems. The challenges are so great that knowledge, paradoxically, can be incapacitating: the more you learn, the harder you find it to believe that a noncatastrophic resolution is conceivable. It is perhaps for this reason that scientists themselves sometimes seem to feel the most optimistic about fields outside their own expertise. The distinguished physicist Freeman Dyson has argued for years that trees should be modified genetically to "convert most of the carbon that they absorb from the atmosphere into some chemically stable form and bury it underground. Or they could convert the carbon into liquid fuels and other useful chemicals."[18] Planting trees is a focus of many environmental action plans because trees draw carbon dioxide from the atmosphere as they grow, and turn most of the carbon into plant material. This is beneficial in the short term, although the results don't last forever, because trees eventually return all that carbon to the atmosphere—when they die and decompose or when they burn in forest fires, woodstoves, or Swedish power plants. (A tree, like a can of Diet Coke, is only a temporary carbon-dioxide sink.) Dyson's proposed solution to this difficulty is to replant "one quarter of the world's forests" with yet-to-be-invented "carbon-eating varieties of the same species," trees genetically modified to sequester carbon by turning it into some form more enduring than wood (perhaps diamonds, one skeptic has suggested[19]). If this is done, Dyson writes, "the forests would be preserved as ecological resources and as habitats for wildlife, and the carbon dioxide in the atmosphere would be reduced by half in about fifty years." This is a biological solution that only a physicist could love. Even if it isn't merely screwy, it's ripe with the potential for unintended consequences—including, not least of all, the eventual necessity of turning *off* all that genetically modified, carbon-annihilating vegetation, to prevent it from reversing the greenhouse effect, which, after all, is the mechanism that makes life on earth possible in the first place.

At that same energy seminar, a common theme among the scientists on the panel was that for several decades We (scientists in general) have been warning You (everyone else) about climate change and other pressing environmental problems, but You and Your politicians haven't heeded Us—and what will it take to get Your attention so that We can get more funding? There was no mention of all the science that went into creating the problems that science would now like to solve if only someone would cough up the dough. The story of the successful and relatively rapid elimination, worldwide, of almost all the industrial uses of chlorofluorocarbons—ozone-destroying chemical compounds, formerly used widely as propellants and refrigerants—was presented as an example of scientists quickly averting a potential global environmental disaster, but there was no mention of the role that scientists had played in creating that same disaster, by inventing chlorofluorocarbons and then steadily extending their industrial applications. Scientists don't get a free pass just because they have a deeper understanding of the various means by which they are helping us do ourselves in. If the gloomiest experts on the panel were remotely accurate in their prognostications, then they and their colleagues have actually let the rest of us down, by failing to take a far more activist public-policy role, worldwide, for decades.

Even among scientists, there is often a tendency to pin the

greatest hopes on untested new ideas, as opposed to relatively well-understood, established ones. This is the same quirk of human nature which leads the owners of professional sports teams to offer vastly larger salaries to promising but inexperienced young draft picks than to proven veterans: with the draft picks anything still seems possible, while the performance potential of the veterans can no longer be viewed as unlimited, even though with them the likelihood of deep disappointment is lower as well. The American decision to launch the ongoing war in Iraq was the product of a similarly unreasonable faith in the untried.* Innumerable exciting energy-related scientific breakthroughs lie ahead of us—almost every morning's newspaper brings news of another—but we already have a good idea of what we need to do, or at least of how to get started.

IN 1973, TWO AMERICAN PROFESSORS OF MATHEMATICS, George B. Dantzig and Thomas L. Saaty, published a book called

---

* Another example: There is a fair amount of interest, among scientists as well as science-fiction buffs, in the concept of "terraforming," or creating an earthlike atmosphere, on Venus, Mars, or some other planet, to make it habitable by humans. Carl Sagan was a proponent (he suggested seeding Venus's atmosphere with algae), and NASA takes the idea seriously, at least when seeking congressional support for manned space travel. But if creating an entire functioning atmosphere, from scratch, on a planet tens of millions of miles away is possible, then slightly tweaking the proportion of a single molecule in earth's own atmosphere, CO_2, ought to be a cinch, right? Terraforming seems feasible, to the extent that it does, only because the challenges are all hypothetical and therefore easy for daydreamers to solve.

Compact City: A Plan for a Liveable Urban Environment.[20] Their radically utopian concept—the Compact City of the title—took the urban-density idea to what can safely be considered its logical extreme. The city they envisioned was a circular, fully enclosed, sixteen-level habitat for two million residents, with a real estate footprint of a little less than nine square miles. Each level was to be built on a reinforced-concrete slab, held up by concrete columns and connected to other levels by elevators and ramps, as in a parking garage. Residents desiring views of the surrounding countryside would be able to opt for (presumably costlier) dwellings situated along the perimeter, facing out. Stores, schools, factories, and other nonresidential uses would be concentrated, mall-style, in the central core. Individual heating and air-conditioning units would be unnecessary because the city itself would be climate-controlled. The domed roof would be made of triangular glass panels and would owe a design debt to R. Buckminster Fuller. "Most houses in Compact City would have two floors in order to conserve base area," the authors wrote, with the self-assurance of professionally logical men who are certain they have thought of everything (Dantzig was an inventor of linear programming). "Design of both the interior and exterior of these houses would vary according to the preferences of the residents. The ringway would provide access to the *rear* of the upper floor of a house. To facilitate home deliveries by electric-battery-powered trucks from the ringway, it would probably be convenient to have the upper floor of a house built to open directly onto the ringway. The lower floor, however, could be offset

10 feet from the ringway 30 feet below, creating an appearance of openness and spaciousness there. See Figure 3-8."[21]

Reading *Compact City* today provides a useful reminder of the vanishingly brief half-life of big ideas. Noble plans to reconfigure the world inevitably run into the world itself. In urban planning in particular, the best, most enduringly fruitful concepts have usually arisen accidentally, and have endured not because anyone was wise enough to identify and preserve them but because they serendipitously developed what was, in effect, a life of their own and were therefore able to withstand the best intentions of potential destroyers, including urban planners themselves. I believe that Manhattan and other truly dense, mixed-use urban areas represent an invaluable template for efficiently arranging a growing global population in a time of shrinking access to a broad range of natural resources, but how to apply that template remains a frustrating mystery, at least to me. Identifying problems is always easier than devising solutions, especially if the problems are big and the solutions are expected to work. One depressing indication of the scale of the world's present environmental difficulties is that there is nothing approaching a consensus about the ideal course of action for individuals, much less for entire nations.

When I first wrote about the environmental advantages of Manhattan-style population density, in *The New Yorker* in 2004,[22] I was often asked why, if I thought urban life was so great, my wife and I hadn't abandoned what I myself had described as our extravagantly wasteful life in the country, and moved back to New York City. This is an obvious question, and a reasonable one—but it highlights a telling flaw in the way most of us tend to think about the environment. If Ann and I left Connecticut tomorrow and moved back into the Manhattan apartment we rented as newlyweds, we might hugely reduce our personal environmental footprint, but we would leave humanity's environmental footprint unchanged, because in order to move we would have to sell our house and our cars and most of the rest of our possessions to other people, who would continue to use them—and life, on balance, would go on as before. The world would be no better off than if we had found a Manhattan family similar to ourselves and simply swapped residences, furniture, and utility bills. The residences, furniture, and utility bills would be in different hands, but they would still exist: there is a difference between changing one's own circumstances and changing the circumstances of the world. As long as two-hundred-plus-year-old houses in small New England towns continue to exist and be inhabited, it's probably not a bad thing for them to be inhabited by people like us, since we work at home and therefore don't have to drive anywhere to work. (One relatively foolproof way for people to shrink their carbon footprints is to work closer to where they live, or live closer to where they work.) Given that reality, what Ann and I really need to do is find ways to reduce the personal energy and carbon profligacy that are built into our existing home and hometown—to apply the Manhattan template to our own lives, to the extent that we can. Like most Americans in the past couple of years, we began to do this as the price of oil soared and the economy imploded. We did it by driving less, turning down our thermostats, cutting our electricity use by more than 20 percent (mainly by getting better at turning things off, and by making far less use of the window air conditioners in our third-floor offices), and by finally getting around to adding more insulation to our attic and basement, a project I'd been meaning to tackle for more than twenty years. We have also realized that we could "live smaller" within our own home, by moving our offices downstairs, into the kids' old bedrooms, and cutting off the heat to the third floor, reconfiguring our living space to make it more compact. It should be noted that these and other (admittedly modest) plans for change were driven not by environmental high-mindedness on our part but by the economic pressure of soaring energy costs—and that several of them were derailed by the subsequent sudden drop in the price of gasoline and home heating oil, in the second half of 2008. (My golf buddies and I were much more diligent about carpooling at $4.80 a gallon than we are at $1.55; I lost momentum on my insulation project when heating oil fell below $3.00; we're both still in our third-floor offices.) Those facts point to an important lesson: environmental solutions that depend solely on willpower are doomed to fail. Plans that are designed, instead, to harness and direct human nature—such as the instinctive human aversion to going broke—are far more likely to succeed, as long as the incentives remain in place.

Even better than financial incentives are living arrangements that, like those in dense cities, force residents to do the right thing automatically, without encountering their conscience or emptying their wallet. Yet no prescription for the environmental salvation of the United States can be viewed as rational if it is based on bulldozing suburban Atlanta and replacing it with domed human Habitrails designed by math professors. We have built our country as we have built it, and we're obviously not going to tear it down and start over. Utopian daydreaming can be useful if it suggests promising initiatives, but it's not a solution in itself: we aren't going to begin again from an empty continent. What we can do, if we can find the will to act, is reduce the many built-in incentives that tend to aggravate our growing environmental difficulties, and increase the incentives that tend to alleviate them. We must also undertake major energy- and climate-related initiatives that are beyond the scope of any economic force smaller than a federal government, including upgrading the American power grid and investing intelligently in the large-scale centralized production of renewable energy, as well as the imposition of taxes, fees, and land-use regulations that reverse the relentless growth of reckless energy consumption. Our current difficulties present an opportunity to use energy-related initiatives to put people back to work, but if we proceed recklessly we will merely compound our existing problems, by wasting stimulus funds on misguided or poorly designed projects. Adding public transit in places that aren't dense enough to support it, expanding our already environmentally disastrous network of roads and highways, and supporting flawed alternative-energy schemes—projects that are high on many politicians' wish lists—might stimulate the economy in the short term but can only set us further behind for the future. A truly green economy, furthermore, is certain to be a smaller economy than the one we are used to, for the simple reason that renewable fuels don't offer the same energy leverage as the stuff we currently burn.

In the decades ahead, we must somehow shift new residential and commercial development away from places where population growth and economic growth exacerbate critical environmental problems and toward places where population growth and economic growth help to relieve them. For American cities, that will mean first understanding and then extending the benefits of population density and the thoughtful mixing of uses, as well as acknowledging that in a dense city the truly important environmental issues are less likely to be things like the carbon footprints of apartment buildings than they are to be old-fashioned quality-of-life concerns like education, culture, crime, street noise, bad smells, resources for the elderly, and the availability of recreational facilities—all of which affect the willingness of people to live in efficient urban cores rather than packing up their children and fleeing to the suburbs. Issues like these can be tough for traditional environmentalists to come to terms with, because they don't feel green: Where are the organic gardens and the backyard compost heaps? But these are critical issues nevertheless. The two things I miss the most about Manhattan are second-run movie theaters and pretty good restaurants, both of which are plentiful in the city and scarce beyond it. If the existence of such things helps to make dense-city living attractive or tolerable to residents and potential residents, then second-run movie theaters and pretty good restaurants are environmental assets. Similarly, recorded books, hands-free cell phones, drive-through Starbucks stores, electronic toll collection, and anything else that makes it easier for suburban residents to tolerate long car commutes are powerful environmental liabilities.

There are few solutions that can be applied everywhere, without regard to local conditions. Tiny, superefficient cars are undoubtedly going to be an environmental necessity in sprawling American suburbs, where driving is unavoidable yet most trips require neither extra horsepower nor three rows of seats; those same cars are an environmental disaster in dense cities and other as yet non-auto-dependent regions, where enticing walkers and transit riders to switch to automobiles would represent a major step backward. Increasing the density of existing urban cores is an environmental necessity because doing so will make such places not only more efficient but also more livable, in all the ways that Jane Jacobs memorably identified; taking steps to increase the density of thinly settled towns and exurban areas would be an environmental negative because almost anyone moving to such a place would, by definition, be coming from someplace denser and therefore less wasteful. Counterintuitive dilemmas abound. The Phoenix metropolitan area is one of the world's premier examples of boneheaded sprawl, but parched deserts are not, a priori, terrible places for human beings to live, since they actually have some environmental advantages, assuming that residents can be prevented from trying to grow grass on sand, and assuming that development can be made sufficiently dense: cooling a building from 110 degrees to 75 degrees requires less energy than heating one to 68 degrees from minus-25; solar exposure in desert regions is often high year-round, making all forms of solar-energy harvesting more productive (although the amount of current generated by photovoltaic panels drops steadily as the surface temperature of the panels rises above about 77 degrees, and high heat increases transmission losses); the daily solar peak in desert areas roughly coincides with the daily peak in human energy demand in those same areas, something that often isn't the case in other environments; and desert development doesn't consume arable land, an important advantage, since almost every expert's plan for the future of civilization involves growing much, much more of one crop or another.

A huge and often unmentioned issue underlying all our ongoing environmental problems is the issue of population. There are too many people in the world, and too many more are on the way. This is an issue that, in the United States, both conservatives and liberals have often seemed eager to avoid—for conservatives, perhaps, because it raises questions about family size, birth control, and abortion, and for liberals because it raises questions about immigration. Every one of the world's environmental problems is made worse by increases in the number of humans, and, most of all, by increases in the number of Americans, since U.S. residents—whether manufactured locally or imported from abroad—have the largest carbon and energy footprints in the world.

Among the deepest challenges we face is the fact that even people who think of themselves as deeply committed to environmental issues often work at cross-purposes to one another and to their own goals: environmentalists tend to focus on defending the places where people aren't rather than on intelligently organizing the places where people are; architects are necessarily concerned with individual buildings rather than with the efficient functioning of entire neighborhoods, cities, or regions; land-use officials and urban planners are hobbled by a regulatory framework that was designed to suit cars; traffic engineers tend to focus on making life better for drivers rather than on making life better in general; politicians are constrained in their ability to pursue long-term goals by the necessity of being elected; venture capitalists oversell supposedly green technologies that have little chance of accomplishing anything beyond enriching their inventors and investors; even the very most promising-seeming ideas have a discouraging history of turning out, in the end, to have been misguided. What's more, some of the staunchest (and most effective) opponents of large-scale renewable energy projects—solar collectors in the Mojave Desert, wind turbines in Nantucket Sound—have been environmentalists.

Globally, we must find ways to use the accelerating urbanization of the human race as a tool for easing environmental catastrophe rather than allowing it to become an amplifier of our many current problems. Unfortunately, prosperity can be even harder to contain than poverty. Automobile dependence is a terminal malady, given any plausible scenario regarding world energy use, yet trying to hold back cars has proved to be about as futile as trying to hold back the common cold. Today, vast cities can arise from nothing in a few years, as in Dubai, and the opportunity for meaningful intervention, even if the will to intervene exists, can pass before the need to act has been recognized.

Further complicating all of this is the fact that the sprawling, wasteful society we Americans have built for ourselves remains extraordinarily appealing, not only to ourselves but also to people all over the world, even as the cost of maintaining it has pushed many Americans to the economic breaking point and beyond. I like finding bargains at the big-box stores that have eviscerated the village green in the sprawling town next to my town (even though I also love the village green), and I love living three minutes from a golf course, and I have never truly regretted moving away from Manhattan. I have a neighbor who does virtually nothing in his very large yard other than mowing it with a riding lawnmower—yet cutting his grass is one of the major recreational satisfactions of his life, an opportunity for reflection and meditation, an escape from telephones and children and e-mail, an activity that, unlike most of his life's other activities, yields immediate, tangible, cumulative results. To pretend that any of these things are unimportant or irrelevant is to sidestep one of the principal barriers between us and any reasonable version of a sustainable future. Every environmentalist, economist, urban planner, politician, and Utopian dreamer needs to come to terms, somehow, with the gleam of pride and satisfaction in the eyes of my Buick-adoring driver in Beijing.

Shortly after my *New Yorker* article about the environmental advantages of Manhattan-style urban density was published, I received a telephone call from one of the magazine's readers, who told me that what I had written had changed her life, and that she was now determined to sell her house, in a sprawling subdivision in California, and move. Would I mind telling her the name of my town—the one with the dirt roads and the wild animals in the yards?—because that's where she'd like to live.

You see, there's the challenge.

1. More Like Manhattan

1. Historical statistics are from the Energy Information Administration of the U.S. Department of Energy, usually listed under "Petroleum Products Supplied by Type," available at: www.eia.doe.gov.

2. Mark Ginsburg and Mark Strauss, "New York City—A Case Study in Density After 9/11," Density Conference, Boston, 2003, http://www .architects.org/emplibrary/C6_b.pdf.

3. American Council for an Energy-Efficient Economy.

4. "Inventory of New York City Greenhouse Gas Emissions 2007," City of New York, Mayor's Office of Operations, Office of Long-Term Planning and Sustainability, http://www.nyc.gov/html/om/pdf/ccp_ report041007.pdf.

5. Our annual mileage, which has gone down since our kids went off to college, is fairly close to the national average. In 2001, the most recent year for which complete national data are available, the average American household consisted of 2.58 people, of whom 1.77 were licensed drivers, and put 21,187 miles on their 1.89 cars. The average American in 2001 traveled 35,244 miles by car—a bigger number because on some trips cars have more than one occupant. These statistics are from

the "2001 National Household Travel Survey," published in 2004 by the Federal High Administration of the U.S. Department of Transportation. You can find a comprehensive "Summary of Travel Trends" here: http://nhts.ornl.gov/2001/pub/STT.pdf.

6. David Owen, "The Dark Side," *The New Yorker*, August 20, 2007, p. 28, http://www.newyorker.com/reporting/2007/08/20/070820fa_fact_owen.

7. Carol D. Leonnig, "Area Tap Water Has Traces of Medicines," *The Washington Post*, March 10, 2008.

8. State population and area figures are from the U.S. Census Bureau, www.quickfacts.census.gov/.

9. The first quotation is from the website of the Mannahatta Project, http://www.wcs.org/sw-high_tech_tools/landscapeecology/mannahatta; the second is from Daniel Denton, an English essayist, as quoted in Edwin G. Burrows and Mike Wallace, *Gotham: A History of New York City to 1898* (New York: Oxford University Press, 1999), p. 3.

10. Nick Paumgarten, "The Mannahatta Project," *The New Yorker*, October 1, 2007. Sanderson has published his findings in a book: *Mannahatta: A Natural History of New York City* (New York: Abrams, 2009).

11. Quoted in Fred R. Shapiro, ed., *The Yale Book of Quotations* (New Haven: Yale University Press, 2006), p. 426.

12. Ben Jervey, *The Big Green Apple: Your Guide to Eco-Friendly Living in New York City* (Guilford, Connecticut: Globe Pequot Press, 2006), p. xi.

13. Jervey, *The Big Green Apple,* p. 39.

14. California Energy Commission, "U.S. Gasoline Per Capita Use by State, 2004" and "U.S. Per Capita Electricity Use by State, 2003," http://www.energy.ca.gov, based on consumption data from the Energy Information Administration and population data from the U.S. Census Bureau; and other sources.

15. Herbert Girardet, *Creating Sustainable Cities* (Dartington, England: Green Books, 1999), quoted in Stephen M. Wheeler and Timothy Beatley, eds., *The Sustainable Urban Development Reader* (London: Routledge, 2004), pp. 125–32. Italics in original.

16. For a brief profile of Girardet, see: http://www.treehugger.com/files/2007/05/herbert_girarde.php.

17. The PlaNYC website can be found at: http://www.nyc.gov/html/planyc2030/html/home/home.shtml.

18. See Diane Cardwell, "Buildings Called Key Source of City's Greenhouse Gases," *The New York Times*, April 11, 2007.

19. The City of New York, "PlaNYC: A Greener, Greater New York," statistics, from 2000–2005, in chapter on climate change, http://www.nyc.gov/html/planyc2030/downloads/pdf/greenyc_climate-change.pdf.

20. Robert Gottlieb, *Forcing the Spring: The Transformation of the American Environmental Movement* (Washington, D.C.: Island Press, 1993), pp. 29–30. Gottlieb writes, "Although he wrote for urban, cosmopolitan publications that allowed him to establish urban support for wilderness protection, Muir was nevertheless especially hostile to urban living."

21. Rachel L. Carson, *Silent Spring* (New York: Houghton Mifflin, 1962).

22. Thomas Jefferson, letter to Dr. Benjamin Rush, April 21, 1803.

23. See Jim Murphy, *An American Plague: The True and Terrifying Story of the Yellow Fever Epidemic of 1793* (New York: Clarion Books, 2003), and J. H. Powell, *Bring Out Your Dead: The Great Plague of Yellow Fever in Philadelphia in 1793* (Philadelphia: University of Pennsylvania Press, 1993). Murphy's book is intended for younger readers, but it is concise and well researched, and it contains many old prints.

24. Thomas Jefferson, letter to Dr. Benjamin Rush, April 21, 1803.

25. Thomas Jefferson, letter to William Short, September 8, 1823.

26. Burrows and Wallace, *Gotham*, p. 359; Frank Bergen Kelly, ed., *Historical Guide to the City of New York* (New York: F. A. Stokes, 1909), p. 95; Robert T. Augustyn and Paul E. Cohen, *Manhattan in Maps: 1527–1995* (New York: Rizzoli, 1997); and other sources.

27. From a contemporary account in the *New York Evening Post,* quoted in John Noble Wilford, "How Epidemics Helped Shape the Modern Metropolis," *The New York Times*, April 15, 2008.

28. Thomas Jefferson, letter to James Madison, December 20, 1787.

29. Hugh Miller, *First Impressions of England and Its People* (Edinburgh: W. P. Nimmo, Hay, & Mitchell, 1847), quoted in Jonathan Bate et al., *The Oxford English Literary History*, vol. 8: *The Victorians:1830–1880* (Oxford: Oxford University Press, 2002), p. 18.

30. Ebenezer Howard, *Garden Cities of To-Morrow* (London: Faber & Faber, 1946; originally published 1902), quoted in Stephen M. Wheeler and Timothy Beatley, eds., *The Sustainable Urban Development Reader* (London: Routledge, 2004), pp. 12–13. Selection from *Garden Cities of To-Morrow* (including Howard's diagrams) available here: http://www.library.cornell.edu/Reps/DOCS/howard.htm.

31. Tom McCarthy, *Auto Mania: Cars, Consumers, and the Environment* (New Haven: Yale University Press, 2007); "6 percent," p. 238; SUV names, p. 234.

32. Anthony Swofford, "A Meditation on Change," *The New York Times*, January 15, 2006.

33. Melissa Holbrook Pierson, *The Place You Love Is Gone* (New York: W. W. Norton, 2006); "lacerating nostalgia," p. 39; "I can't help it," p. 43; "water closets," p. 126; "condemning," p. 133.

34. Pierson, *The Place You Love Is Gone*, p. 164.

35. Jane Jacobs, *The Death and Life of Great American Cities* (New York: Vintage Books, 1992; originally published 1961).

36. Jacobs, *Death and Life*, p. 4.

37. Burrows and Wallace, *Gotham*.

38. Quoted in Augustyn and Cohen, *Manhattan in Maps: 1527–1995*, p. 102.

39. Burrows and Wallace, *Gotham*, p. 421.

40. Jacobs, *Death and Life*, p. 257.

41. John Holtzclaw, from "Convenient Neighborhood, Skip the Car," a talk presented at the ninety-third annual meeting and exhibition of the Air & Waste Management Association, Salt Lake City, Utah, June 23, 2000.

42. The definitive biography of Moses is Robert A. Caro's, which won the Pulitzer Prize: *The Power Broker: The Rise and Fall of New York* (New York: Vintage Books, 1975).

43. John Nolen, *Twenty Years of City Planning Progress in the United States* (National Conference on City Planning, 1927), p. 12, quoted in Daniel Lazare, *America's Undeclared War* (New York: Harcourt, 2001), pp. 117–18.

44. Clifford Krauss, "Can They Really Call the Chainsaw Eco-Friendly?" *The New York Times*, June 25, 2007.

45. Brian Stelter, "A Network to Make an Environmental Point," *The New York Times*, June 2, 2008.

46. "Transylvania Commissioners Sink Teeth into Green Jail Plan" was the headline of another article in the same issue. The magazine's website is: www.correctionalnews.com.

47. For your own BE GREEN bumper sticker, send a stamped, self-addressed envelope to BE GREEN, 564 Gateway Drive, Napa, CA 94558. From the macaroni box: "As the stewards of our fragile planet, we humans need to unite and speak out on behalf of all of Earth's inhabitants—from plankton to polar bears to whales to redwoods. We are all interconnected. We all share the same home. Displaying a BE GREEN sticker gives you this voice."

48. Ted Nordhaus and Michael Shellenberger, *Break Through: From the Death of Environmentalism to the Politics of Possibility* (Boston: Houghton Mifflin, 2007), pp. 113–14.

49. Doug Struck, "Canada Alters Course on Kyoto," *The Washington Post*, May 3, 2006.

2. Liquid Civilization

1. Neela Banerjee, "Many Feeling Pinch After Newest Surge in U.S. Fuel Prices," *The New York Times*, June 1, 2004.

2. James Finch, "Ethanol, Fertilizer and Higher Natural Gas Prices," *The Market Oracle*, April 29, 2007, http://www.marketoracle.co.uk/Article891.html. See also statistics compiled by the International Fertilizer Industry Organization, http://www.fertilizer.org/ifa/statistics/indicators/ind_reserves.asp.

3. "20 percent" from Robert Bryce, *Gusher of Lies: The Dangerous Delusions of "Energy Independence"* (New York: PublicAffairs, 2008), p. 65. "Rise to a third" is a USDA forecast, from Scott Kilman, "Corn Prices Will Remain in Record High Territory," *The Wall Street Journal*, May 10–11, 2008. The fuel-consumption arithmetic is mine, based on statistics from EIA. For more on Brazilian ethanol, see Bryce, *Gusher of Lies*, pp. 167–75.

4. Jeffrey D. Sachs, "Surging Food Prices and Global Stability," *Scientific American*, June 2008.

5. Bob Davis and Douglas Belkin, "Food Inflation, Riots Spark Worries for World Leaders," *The Wall Street Journal,* April 14, 2008.

6. Keith Bradsher and Andrew Martin, "Hoarding Nations Drive Food Costs Ever Higher," *The New York Times,* June 30, 2008.

7. For more on peak oil, see Kenneth S. Deffeyes, *Hubbert's Peak: The Impending World Oil Shortage* (Princeton, New Jersey: Princeton University Press, 2001) and any of a number of websites, among them www.hubbertpeak.com and www.peakoil.com, as well as numerous other sources.

8. David Goodstein, *Out of Gas: The End of the Age of Oil* (New York: W. W. Norton, 2004), pp. 15, 123.

9. "Drowning in Oil" and "Why Cheap Oil May Be Bad," *The Economist,* March 4, 1999.

10. Bryce, *Gusher of Lies,* pp. 36–37. Bryce's prose is hyperbolic, and he misidentifies uranium as a fossil fuel, but his book is an excellent corrective for anyone who believes that the widely embraced idea of seeking to make the United States independent of reliance on foreign fossil fuels is remotely plausible or sensible.

11. Jared Diamond, *Collapse: How Societies Choose to Fail or Succeed* (New York: Viking, 2005), p. 114.

12. Steven D. Levitt and Stephen J. Dubner, *Freakonomics: A Rogue Economist Explores the Hidden Side of Everything* (New York: HarperCollins, 2006), pp. 268–69.

13. Jeffrey S. Dukes, "Burning Buried Sunshine: Human Consumption of Ancient Solar Energy," March 25, 2003, http://globalecology.stanford .edu/DGE/Dukes/Dukes_ClimChange1.pdf. The original study was published in the November 2003 issue of *Climatic Change.* Dukes, who is now on the faculty of Purdue University, made the wheat-field analogy in an interview published by the University of Utah in a press release dated October 23, 2003, http://www.eurekalert.org/pub_ releases/2003-10/uou-bm9102603.php.

14. For more on the abiogenic theory of fossil-fuel formation, see David Osborne, "The Origin of Petroleum," *The Atlantic Monthly,* February 1986, and Thomas Gold, *The Deep Hot Biosphere* (New York: Copernicus Books, 1999).

15. David Owen, *Sheetrock & Shellac: A Thinking Person's Guide to the Art*

and Science of Home Improvement (New York: Simon & Schuster, 2007), pp. 162–63.

16. For more, see the website of the U.S. Department of Energy's Energy Information Agency, http://www.eia.doe.gov/oiaf/ieo/coal.html.

17. For more, see http://www.njskylands.com/hsoil.htm.

18. *California Fireside Journal*, September 3, 1860.

19. Tom McCarthy, *Auto Mania: Cars, Consumers, and the Environment* (New Haven: Yale University Press, 2007), pp. 16–17.

20. "Lightning Reveals Oil in the Town of Olean," *The New York Times*, July 25, 1920.

21. *Homemade Contrivances and How to Make Them: 1001 Labor-Saving Devices for Farm, Garden, Dairy, and Workshop* (New York: Skyhorse, 2007; originally published by Orange Judd of New York in 1897).

22. The 12,000-watt figure comes from Elizabeth Kolbert, "The Island in the Wind," *The New Yorker*, July 7 and 14, 2008.

23. Loren Eiseley, *The Star Thrower* (New York: Harvest Books, 1979), p. 45.

24. Goodstein, *Out of Gas*, p. 32.

25. Dick Cheney quoted in John Cassidy, "Pump Dreams," *The New Yorker*, October 11, 2004.

26. For more on this complicated subject, see Bryce, *Gusher of Lies*. Regarding the cost-ineffectiveness of ethanol, Robert Hahn, who was the cochair of the U.S. Alternative Fuel Council under the first President Bush, has written: "If annual production increases by three billion gallons in 2012—a plausibly modest number when the EPA made its own calculations—we estimate that the costs will exceed the benefits by about $1 billion a year. If domestic production reaches the more 'optimistic' Energy Department projection for that year, net economic costs would likely top $2 billion annually." Hahn's essay "Ethanol's Bottom Line" appeared in *The Wall Street Journal*, November 24, 2007.

27. For a concise introduction to the extreme difficulty of economically powering cars with hydrogen, see Matthew L. Wald, "Questions About a Hydrogen Economy," *Scientific American*, May 2004.

28. Romm quoted in Robert S. Boyd, "Hydrogen Cars May Be a Long

Time Coming," *McClatchy Newspapers*, May 15, 2007. Joseph J. Romm, *The Hype about Hydrogen: Fact and Fiction in the Race to Save the Climate* (Washington, D.C.: Island Press, 2005).

29. Thomas L. Friedman, "The Democratic Recession," *The New York Times*, May 7, 2008.

30. World Commission on Environment and Development, *Our Common Future* (Oxford: Oxford University Press, 1987), p. 45.

31. Russell Gold, "As Prices Surge, Oil Giants Turn Sludge into Gold," *The Wall Street Journal*, March 27, 2006.

32. The survey was conducted online, and the sample was small—just 501 respondents—and the project was partly sponsored by Archer Daniels Midland, which is not only involved in the manufacture of bioplastics but also bears a major responsibility for the economic distortions built into the U.S. corn market and for the absurd federal subsidies for the production of ethanol, but the results are consistent with my own informal sampling. You can find a press release here: http://www.metabolix.com/publications/pressreleases/PR20070420 .html.

33. Stephen Fenichell, *Plastic: The Making of a Synthetic Century* (New York: Harper Business, 1996), p. 4.

34. Natalie Angier, "Adored, Deplored and Ubiquitous," *The New York Times*, April 15, 2008.

35. U.S. Energy Information Administration, "Petroleum Products: Supply," www.eia.doe.gov/neic/infosheets/petroleumproducts.html.

36. Charles Moore, "Across the Pacific Ocean, Plastics, Plastics, Everywhere," *Natural History*, November 2003.

37. For more on the sick lobsters, see: http://www.mbl.edu/news/press_releases/2008_pr_08_13.html.

38. John De Graaf, David Wann, and Thomas H. Naylor, *Affluenza: The All-Consuming Epidemic* (New York: Berrett-Koehler, 2001), p. 85. The authors say the figures are from the United Nations Environment Programme, http://www.unep.org/.

39. Bryce, *Gusher of Lies*, p. 175.

40. U.S. Patent No. 226945, "Automobile Body Construction," filed by H. Ford, May 15, 1940, issued January 13, 1942.

41. David Owen, "The Soul of a New Dessert," *Harper's*, October 1983.

42. "We Can Do More," *Scientific American*, August 2008.

43. Peter Rogers, "Facing the Freshwater Crisis," *Scientific American*, August 2008.

44. The Pentagon's energy figures are from Sohbet Karbuz, "US Energy Consumption—Facts and Figures," *Energy Bulletin*, May 20, 2007, and are footnoted there, http://www.energybulletin.net/node/29925.

45. Patricia Marx, "Buy Shanghai!" *The New Yorker*, July 21, 2008.

46. Energy Information Administration, U.S. Department of Energy, http://www.eia.doe.gov/pub/oil_gas/petroleum/analysis_publications/oil_market_basics/demand_text.htm.

47. Robert Reich interviewed in "Short-Straw Economics," *The New York Times Magazine*, July 8, 2008.

48. William Stanley Jevons, *The Coal Question: An Inquiry Concerning the Progress of the Nation, and the Probable Exhaustion of Our Coal-Mines* (London: Macmillan, 1865). The complete text of the book's second edition, published in 1866, is available online: http://www.econlib.org/library/YPDBooks/Jevons/jvnCQ.html.

49. Peter Huber and Mark P. Mills, *The Bottomless Well: The Twilight of Fuel, the Virtue of Waste, and Why We Will Never Run Out of Energy* (New York: Basic Books, 2006), p. 111.

50. Kolbert, "The Island in the Wind."

51. Scott Horsley, "Gas Prices Continue to Soar," National Public Radio, March 11, 2008, http://www.npr.org/templates/story/story.php?storyId=88116151.

52. The *Simpsons* episode is titled "Homer to the Max." It was written by John Swartzwelder and first aired in 1999.

3. There and Back

1. Tom McCarthy, *Auto Mania: Cars, Consumers, and the Environment* (New Haven: Yale University Press, 2007); re: farms and small towns, p. 37; re: second cars, pp. 149–51; "only 3 percent," p. 148.

2. Federal Highway Administration, U.S. Department of Transportation, "Summary of Travel Trends," 2004, http://nhts.ornl.gov/2001/pub/STT.pdf.

3. Energy Information Agency, U.S. Department of Energy, www.eia.doe

.gov. The 1920 figure is extrapolated from "Oil Production in 1920," *The New York Times*, March 13, 1921, and from historical data compiled by the EIA, which has old Bureau of Mines records going back to 1918.

4. Richard M. Haughey, *Higher-Density Development: Myth and Fact* (Washington, D.C.: Urban Land Institute, 2005), p. 10. The statistics about schools in Minneapolis/St. Paul are from Brett Hulsey, *Sprawl Costs Us All* (Madison, Wisconsin: Sierra Club Midwest Office, 1996).

5. Speaking of compact fluorescent lamps (CFLs), have you seen the U.S. Environmental Protection Agency's recommendation for what to do if one of them breaks? Here are highlights:

> *Because CFLs contain a small amount of mercury, EPA recommends the following clean-up and disposal guidelines:*
>
> *1. Before Clean-up: Ventilate the Room*
> - *Have people and pets leave the room, and don't let anyone walk through the breakage area on their way out.*
> - *Open a window and leave the room for 15 minutes or more.*
> - *Shut off the central forced-air heating/air conditioning system, if you have one.*
>
> *2. Clean-Up Steps for Hard Surfaces*
> - *Carefully scoop up glass fragments and powder using stiff paper or cardboard and place them in a glass jar with metal lid (such as a canning jar) or in a sealed plastic bag.*
> - *Use sticky tape, such as duct tape, to pick up any remaining small glass fragments and powder.*
> - *Wipe the area clean with damp paper towels or disposable wet wipes and place them in the glass jar or plastic bag.*
> - *Do not use a vacuum or broom to clean up the broken bulb on hard surfaces.*

That's just the beginning. For the complete protocol, see: http://energystar.custhelp.com/cgi-bin/energystar.cfg/php/enduser/std_adp.php?p_faqid=2655.

6. Henry Ford, *Ford Ideals: Being a Selection from "Mr Ford's Page" in the Dearborn Independent* (Dearborn, Michigan: Dearborn, 1922), p. 157.

7. Michael Pollan, "Why Bother?" *The New York Times Magazine*, April 20, 2008.

8. The Gordon Strong Automobile Objective was among several Wright projects featured in "Frank Lloyd Wright: Designs for an American Landscape, 1922–1932," an exhibition at the Library of Congress in late 1996 and early 1997. You can find images and other information here: http://www.loc.gov/exhibits/flw/flw02.html. For an extensive description of the project, see Mark Reinberger, "The Sugarloaf Mountain Project and Frank Lloyd Wright's Vision of a New World," *Journal of the Society of Architectural Historians*, March 1984. Also of interest is the website for Sugarloaf Mountain, which today is owned by Stronghold, Inc., a nonprofit corporation established by Gordon Strong in 1946. The Sugarloaf website (www.sugarloafmd.com) makes no mention of the Automobile Objective or of Wright.

9. Frank Lloyd Wright, letter to Gordon Strong, October 20, 1925, quoted in Reinberger, "The Sugarloaf Mountain Project and Frank Lloyd Wright's Vision of a New World," p. 46.

10. Frank Lloyd Wright, *The Living City* (New York: Horizon Press, 1958). Quotations are from a 1970 reprint, published by Plume Books. The two earlier books subsumed by *The Living City* were *The Disappearing City* (New York: W. F. Payson, 1932) and *When Democracy Builds* (Chicago: University of Chicago Press, 1945).

11. Wright, *The Living City*, p. 33.

12. Wright, *The Living City*, p. 59.

13. Wright, *The Living City*, p. 61.

14. Peter Newman and Jeffrey Kenworthy, *Sustainability and Cities: Overcoming Automobile Dependence* (Washington, D.C.: Island Press, 1999), p. 31.

15. Herbert Hoover, *Zoning Primer* (Washington, D.C.: Department of Commerce, 1922), p. 1, quoted in Joseph P. Schwieterman and Dana M. Caspall, *The Politics of Place: A History of Zoning in Chicago* (Chicago: Lake Claremont Press, 2006), p. 25. The constitutionality of zoning ordinances was upheld by the U.S. Supreme Court in 1926, in *Village of Euclid, Ohio, et al. v. Ambler Realty Company*. At issue was a contention by Ambler, a developer, that the town of Euclid, a suburb of Cleveland, had illegally prevented it from developing a particular 68-acre

parcel as an industrial site. The court ruled that Euclid's zoning regulations, which classified part of the parcel as residential and therefore required the entire parcel to meet the stricter requirements governing that zone, were not arbitrary and that Ambler had not been denied due process, and it established the overall principle that zoning regulations are "a proper exercise of the police power" and a legitimate extension of any municipality's right to control nuisances. That decision, which has never been seriously challenged, helped to fuel the rapid spread of similar development controls all over the country.

16. "The Real Price of Gasoline: Report No. 3, An Analysis of the Hidden External Costs Consumers Pay to Fuel Their Automobiles," International Center for Technology Assessment, Washington, D.C., 1998.

17. See Nick Paumgarten, "There and Back Again," *The New Yorker*, April 16, 2007.

18. Journey-to-work statistics for 1960 through 2000 are from the U.S. Census Bureau and the Bureau of Transportation Statistics of the U.S. Department of Transportation.

19. Clifford Krauss, "Gas Prices Send Surge of Riders to Mass Transit," *The New York Times*, May 10, 2008.

20. Paul Krugman, "Stranded in Suburbia," *The New York Times*, May 19, 2008.

21. New York City Department of Transportation, quoted in William Neuman, "The City's Boom Years Were Good for Transit, Too," *The New York Times*, December 14, 2008.

22. Metropolitan Transit Authority, http://www.mta.info/nyct/facts/ffsubway.htm. New York City's subway system carried 1.5 billion passengers in 2006.

23. Metropolitan Transit Authority, http://www.mta.info/nyct/facts/ffsubway.htm. I've combined the passenger figures for New York's Transit Authority (740 million) and Department of Transportation (102 million), both of which run buses in the city.

24. Boris S. Pushkarev and Jeffrey M. Zupan, *Public Transportation and Land Use Policy* (Bloomington: University of Indiana Press, 1977).

25. Boris S. Pushkarev and Jeffrey M. Zupan, "Urban Rail in America: An Exploration of Criteria for Fixed Guideway Transit," Urban Mass Transportation Administration, U.S. Department of Transportation, 1980;

John Holtzclaw, "Explaining Urban Density and Transit Impacts on Auto Use," Natural Resources Defense Council, 1991.

26. Newman and Kenworthy, *Sustainability and Cities*, p. 87.

27. John Holtzclaw, "Does a Mile in a Car Equal a Mile on a Train? Exploring Public Transit's Effectiveness in Reducing Driving," Sierra Club, http://www.sierraclub.org/sprawl/articles/reducedriving.asp.

28. Bill McKibben, *Hope, Human and Wild: True Stories of Living Lightly on the Earth* (Minneapolis: Milkweed Editions, 2007), pp. 68–69. This is a reprint edition of the book, which was published originally in 1995.

29. Richard Register, *Ecocities: Rebuilding Cities in Balance with Nature* (Gabriola Island, B.C., Canada: New Society Publishers, 2006), pp. 316–17.

30. Newman and Kenworthy, *Sustainability and Cities*, pp. 30–31.

31. Sharon Feigon and David Hoyt, "Land Use and Greenhouse Gas Emissions from Transportation," Center for Neighborhood Technology, 2003.

32. Ignacio San Martín, "The Character of Urban Sprawl and Its Indicators," First Policy Forum on Urban Sprawl, Center for Environmental Studies, Budapest, Hungary, 2000.

33. The sprawling development of Phoenix is aptly described by the caption below a photograph on a NASA website: "Phoenix is actually surrounded by 22 towns and cities that have grown so closely together in recent decades it is almost impossible to distinguish one from another in this 30-meter-resolution satellite image." The satellite photograph can be found at: http://earthobservatory.nasa.gov/Newsroom/NewImages/images.php3?img_id=15282. The photograph was taken in 2002. Phoenix's horizontal growth has, if anything, accelerated since then.

34. Jane Jacobs, *The Death and Life of Great American Cities* (New York: Vintage Books, 1992; originally published 1961), p. 352.

35. M. A. Farber, "Negotiations with Post Keep Kheel in Limelight," *The New York Times*, February 20, 1988.

36. Stanley Levey, "Autos Aggravate Transit Problem; Kheel Seeks Cure," *The New York Times*, May 31, 1955.

37. Richard Oliver, "Kheel Proposes Tolls on Major Highways," *New York Daily News*, October 10, 1969.

38. Jacobs, *The Death and Life of Great American Cities*, pp. 338ff.

39. Ian Urbina, "F.D.R. Drive Taking Its Act to the River, for Three Years," *The New York Times*, May 18, 2004.

40. Robert Cervero, *The Transit Metropolis: A Global Inquiry* (Washington, D.C., 1998), p. 22.

41. Ariel Hart, "HOV Lane Fees Could Start by 2010," *Atlanta Journal-Constitution*, March 20, 2008.

42. Tom Vanderbilt, *Traffic: Why We Drive the Way We Do (and What It Says About Us)* (New York: Alfred A. Knopf, 2008), p. 114.

43. Timothy Beatley, *Green Urbanism: Learning from European Cities* (Washington, D.C.: Island Press, 2000), p. 164.

44. During the Minnesota study, in 2000, 433 meters were turned off for six weeks, and the resulting traffic patterns were studied by Cambridge Systematics. A copy of the study summary can be found at: http://www.dot.state.mn.us/rampmeterstudy/pdf/execsummary/executivesummary.pdf.

45. The website of Transportation Alternatives can be found at: www.transalt.org.

46. Newman and Kenworthy, *Sustainability and Cities*, p. 183.

47. "A Bolder Plan: Balancing Free Transit and Congestion Pricing in New York City," Nurture New York's Nature, Inc. A digital copy of the proposal can be found at: www.nnyn.org/kheelplan.

48. Ori Brafman and Rom Brafman, *Sway: The Irresistible Pull of Irrational Behavior* (New York: Currency/Doubleday, 2008), pp. 48–49. For the SoBe study, see Baba Shif, Ziv Carmon, and Dan Ariely, "Placebo Effects of Marketing Actions: Consumers May Get What They Pay For," *Journal of Marketing Research*, November 2005, http://www.predictablyirrational.com/pdfs/Placebo1.pdf.

49. "A Bolder Plan," pp. 14–15, 24.

50. Jeff Sabatini, "Daimler's Minicar: More Charming Than Smart," *The Wall Street Journal*, March 21, 2008.

51. Eric A. Taub, "Ready for Its Hollywood Close-Up," *The New York Times*, May 11, 2008.

52. The City Car website is: http://cities.media.mit.edu/projects/citycar.html, and additional information about the project can be found on other MIT websites.

4. The Great Outdoors

1. Metropolitan Transit Authority, http://www.mta.info/nyct/facts/ffsubway.htm.
2. Dashka Slater, "Walk the Walk," *The New York Times Magazine*, April 20, 2008.
3. Clive Thompson, "Why New Yorkers Last Longer," *New York*, October 6, 2007.
4. Eleanor M. Simonsick et al., "Measuring Higher Level Physical Function in Well-Functioning Older Adults," *Journal of Gerontology: Medical Sciences*, October 2001, pp. 644–49.
5. Slater, "Walk the Walk."
6. Ben Harder, "Weighing In on City Planning," *Science News*, January 20, 2007.
7. Jane Jacobs, *The Death and Life of Great American Cities* (New York: Vintage Books, 1992; originally published 1961), p. 259.
8. Jacobs, *Death and Life*, pp. 265–66.
9. John Holtzclaw, "Curbing Sprawl to Curb Global Warming," Sierra Club, http://www.sierraclub.org/sprawl/articles/warming.asp.
10. Douglas Farr, *Sustainable Urbanism: Urban Design with Nature* (Hoboken: John Wiley & Sons, 2008), p. 21.
11. Calvin Trillin, "Rudy Giuliani, Proctor of New York," *Time*, March 2, 1998.
12. See Peter Newman and Jeffrey Kenworthy, *Sustainability and Cities: Overcoming Automobile Dependence* (Washington, D.C.: Island Press, 1999), p. 147 and elsewhere. There is also much useful information about traffic calming here: http://www.trafficcalming.org/.
13. Christopher Alexander et al., *A Pattern Language: Towns, Buildings, Construction* (New York: Oxford University Press, 1977), p. 271.
14. William Neuman and Fernanda Santos, "On 3 Days in August, City Will Try No-Car Zone," *The New York Times*, June 17, 2008.
15. Cristina Milesi et al., "Mapping and Modeling the Biogeochemical Cycling of Turf Grasses in the United States," *Environmental Management*, September 2005, pp. 426–38.
16. Elizabeth Kolbert, "Turf War," *The New Yorker*, July 21, 2008.
17. Richard Louv, "Leave No Child Inside," *Orion*, March/April, 2007, available here: http://www.orionmagazine.org/index.php/articles/article/240/.

18. David Biello, "Are Americans Afraid of the Outdoors?" *Scientific American*, February 5, 2008. A brief version of this article appeared in the April 2008 issue of the magazine; a longer version appears online at: www.sciam.com.

19. François Leydet, *The Last Redwoods and the Parkland of Redwood Creek* (San Francisco: Sierra Club, 1963), p. 132, quoted in Ted Nordhaus and Michael Shellenberger, *Break Through: From the Death of Environmentalism to the Politics of Possibility* (Boston: Houghton Mifflin, 2007), p. 26.

20. Statistics from the National Park Service, quoted in "Environmental Awareness," *The Economist*, February 8, 2007. "No park, it seems, is immune to the decline: even in Yosemite, one of the system's oldest parks and probably its best known, the number of visitors dropped 17% over the ten-year period. The number of visitors to Death Valley, an easy drive from vigorously growing Las Vegas, went down 28% over the same span. In some of the system's remoter parks, such as Lava Beds National Monument near the California-Oregon border, site of much fighting in an Indian war of 1872–73, the number of daily visitors is down to ten or fewer."

21. Oliver R. W. Pergams and Patricia A. Zaradic, "Is Love of Nature in the US Becoming Love of Electronic Media? 16-Year-Downtrend in National Park Visits Explained by Watching Movies, Playing Video Games, Internet Use, and Oil Prices," *Journal of Environmental Management*, 80 (2006), pp. 387–93; paper available online at: www .videophilia.org.

22. "Do People Still Care about Nature?" Interview available online at: www.nature.org.

23. Oliver R. W. Pergams and Patricia A. Zaradic, "Evidence for a Fundamental and Pervasive Shift away from Nature-Based Recreaction," *Proceedings of the National Academy of Sciences*, 105:2295–230.

24. June Fletcher, "Giving Up on the Outdoors," *The Wall Street Journal*, June 8, 2007.

25. Jeanne E. Arnold and Ursula A. Lang, "Changing American Home Life: Trends in Leisure and Storage among Middle Class Families," *Journal of Family and Economic Issues*, March 2007, pp. 23–48.

26. Richard Louv, *Last Child in the Woods: Saving Our Children from Nature-Deficit Disorder* (New York: Workman, 2008). The camp we

attended was Big Spring Ranch, in Florissant, Colorado. There's a girls' camp, too, called High Trails Ranch. They are run by the same people who ran them when we were kids. Here's a link to the website: http://www.sanbornwesterncamps.com/.

27. Louv, "Leave No Child Inside."

5. Embodied Efficiency

1. David Gissen, ed., *Big & Green: Toward Sustainable Architecture in the 21st Century* (New York: Princeton Architectural Press, 2002).
2. Daniel Kaplan, "Manhattan's Green Giant," *Environmental Design & Construction*, September 1997.
3. Craig Lambert, "The Hydrogen-Powered Future," *Harvard Magazine*, January/February 2004.
4. More information about RMI can be found here: http://www.rmi.org/.
5. Gissen, *Big & Green*, p. 10.
6. Michael Phillips and Robert Gnaizda, "New Age Doctrine Is Out to Lunch on Three Issues," *CoEvolution Quarterly*, Summer 1980.
7. John Holtzclaw, "Compact Growth," http://www.sierraclub.org/sprawl/community/compact.asp.
8. Casey Logan, "Screwed by Sprint," *The Pitch*, November 21, 2002, http://www.pitch.com/2002-11-21/news/screwed-by-sprint/1.
9. Shirley Christian, "Sprint Is Building Huge Headquarters in Kansas," *The New York Times*, July 12, 1998.
10. Logan, "Screwed by Sprint."
11. Sierra Club, "1998 Sierra Club Sprawl Report: 30 Most Sprawl-Threatened Cities," http://www.sierraclub.org/sprawl/report98/kansas_city.asp.
12. Christian, "Sprint Is Building Huge Headquarters in Kansas."
13. For more on the steps that colleges are taking to enhance public perceptions of their environmental responsibility, see Tracy Jan, "Not to Be Out-Greened," *The Boston Globe*, July 29, 2008.
14. Auden Schendler and Randy Udall, "LEED Is Broken; Let's Fix It," *Grist*, October 26, 2005. A link to the article: http://www.aspensnowmass.com/environment/images/LEEDisBroken.pdf.
15. The Energy Star website is: http://www.energystar.gov/.
16. The study was conducted by CoStar Group, a research company that

gathers and sells information about commercial real estate in the United States and the United Kingdom. A press release regarding the study can be found here: http://www.costar.com/News/Article.aspx?id=D968F1 E0DCF73712B03A099E0E99C679.

17. Anya Kamenetz, "The Green Standard?" *Fast Company*, October 2007.

18. A case study of the Merrill Center project can be found on the U.S. Green Building Council's website: http://leedcasestudies.usgbc .org/overview.cfm?ProjectID=69. Additional information about the building can be found here: http://www.cbe.berkeley.edu/mixedmode/ chesapeake.html. The website of the Chesapeake Bay Foundation is: http://www.cbf.org. In 2007, according to the Public Policy Institute of New York State, the average price of electricity in Maryland was ten or eleven cents per kilowatt-hour: http://www.ppinys.org/reports/jtf/ electricprices.html.

19. The Green Globes website is: http://www.greenglobes.com/.

20. Schendler and Udall, "LEED Is Broken; Let's Fix It."

21. Matt Tyrnauer, "Industrial Revolution, Take Two," *Vanity Fair*, May 2008.

22. Roger Rosenblatt, "The Man Who Wants Buildings to Love Kids," *Time*, February 15, 1999.

23. June Fletcher, "The Price of Going Green," *The Wall Street Journal*, February 29, 2008.

24. Mimi Read, "In Chicago, Tinted Green," *The New York Times*, March 13, 2008.

25. Jenny Higgons, "This House in Pelham Is a Model of Efficiency," *The Journal News*, June 28, 2008, http://www.lohud.com/apps/pbcs.dll/ article?AID=2008806280308. The Ellenbogens received a $7,000 credit on their federal income tax return and a $40,000 rebate from the New York State Energy Research and Development Authority.

26. Michael Moyer and Amanda Schupak, "American Power," *Popular Science*, November 2008.

27. Severin Borenstein, "The Value and Cost of Solar Photovoltaic Electricity Production," Center for the Study of Energy Markets Working Paper Series," University of California Energy Institute, January 2008, p. 22.

28. You can see the *Wired* green house at: http://www.wired.com/promo/ wiredlivinghome/.
29. U.S. Department of Energy, "Why Is Window Area so Important to Energy Code Compliance?" http://resourcecenter.pnl.gov/cocoon/ morf/ResourceCenter/article/11.
30. National Fenestration Rating Council, "Fenestration Heat Loss Facts," www.nfrc.org.
31. Thomas L. Friedman, *Hot, Flat, and Crowded: Why We Need a Green Revolution—and How It Can Renew America* (New York: Farrar, Straus and Giroux, 2008), p. 267.
32. Amory Lovins, *Small Is Profitable: The Hidden Economic Benefits of Making Electrical Resources the Right Size* (Snowmass, Colorado: Rocky Mountain Institute, 2002).
33. Nathanael Greene and Roel Hammerschlag, "Small and Clean Is Beautiful: Exploring the Emissions from Distributed Generation and Pollution Prevention Policies," *Electricity Journal*, June 2000.

6. The Shape of Things to Come

1. *BP Statistical Review of World Energy 2007*, various documents from the Energy Information Agency of the U.S. Department of Energy, and various other sources, among them the website nationmaster.com. The United States is actually just the world's fifteenth-largest per-capita consumer of oil, after the Virgin Islands (nearly a barrel per person per day, or fourteen times the overall U.S. rate), Netherlands Antilles, Singapore, Luxembourg, Kuwait, Qatar, Guam, Faroe Islands, United Arab Emirates, the Bahamas, Saudi Arabia, Bermuda, Canada, and Aruba (2004 figures). The average citizen of the Democratic Republic of the Congo, which ranks 192nd in per-capita oil consumption, uses about a fifth of an ounce of oil per day. See: http://www.nationmaster.com/graph/ene_oil_con_percap-energy-oil-consumption-per-capita.
2. Tom Vanderbilt, *Traffic: Why We Drive the Way We Do (and What It Says About Us)* (New York: Alfred A. Knopf, 2008), p. 218.
3. Edward H. Ziegler, "China's Cities, Globalization, and Sustainable Development: Comparative Thoughts on Urban Planning, Energy, and

Environmental Policy," *Washington University Global Studies Law Review*, vol. 5, 2006, p. 297. Ziegler is a professor at the University of Denver's Sturm College of Law. His specialty is land use.

4. Jeffrey D. Sachs, "Coping with a Persistent Oil Crisis," *Scientific American*, October 2008.

5. Charles Dickens, *Bleak House* (New York: Modern Library Edition, 2002), p. 28.

6. There are some good *hutong* photographs here: http://beijingman.blogspot.com/2007/12/beijing-hutong-time-out.html. See also Wang Wenbo, *Recollections of Hutong* (Beijing: China Nationality Art Photograph Publishing House, 2006)—a wonderful book, if you can find it.

7. Peter Hessler, "Hutong Karma," *The New Yorker*, February 13 and 20, 2006.

8. Stewart Brand, "City Planet," *Strategy + Business*, Spring 2006: http://www.strategy-business.com/press/16635507/06109.

9. Elisabeth Rosenthal, "New Jungles Prompt a Debate on Saving Primeval Rain Forests," *The New York Times*, January 30, 2009.

10. Danielle Pergament, "36 Hours: Dubai," *The New York Times*, April 6, 2008.

11. Seth Sherman, "Dubai, Where Too Much Is Never Enough," *The New York Times*, June 4, 2006.

12. William McDonough and Michael Braungart, *Cradle to Cradle: Remaking the Way We Make Things* (New York: North Point Press, 2002).

13. Jennifer Conlin, "Going Green in Australia's Blue Mountains," *The New York Times*, April 6, 2008.

14. Michael Pollan, *The Omnivore's Dilemma: A Natural History of Four Meals* (New York: Penguin Books, 2006), p. 242.

15. Michael Specter, "Big Foot," *The New Yorker*, February 25, 2008.

16. From "Tesco, Carbon and the Consumer," a speech by Sir Terry Leahy, Tesco's CEO, given in London on January 1, 2007. For Leahy's complete speech, see: http://www.tesco.com/climatechange/speech.asp.

17. Despommier's website is http://www.verticalfarm.com. See also Bina Venkataraman, "Country, the City Version. Farms in the Sky Gain New Interest," *The New York Times*, July 15, 2008.

18. Freeman Dyson, "The Question of Global Warming," *The New York Review of Books*, June 12, 2008.

19. For the comment about diamonds, see: http://www.realclimate.org/index.php/archives/2008/05/freeman-dysons-selective-vision/.

20. George B. Dantzig and Thomas L. Saaty, *Compact City: A Plan for a Liveable Urban Environment* (San Francisco: W. H. Freeman, 1973).

21. Dantzig and Saaty, *Compact City*, p. 44.

22. David Owen, "Green Manhattan," *The New Yorker*, October 18, 2004.

INDEX